KU-571-831

STATISTICS

STATISTICS

Alan Graham

TEACH YOURSELF BOOKS

For UK orders: please contact Bookpoint Ltd, 130 Milton Park, Abingdon, Oxon OX14 4SB. Telephone: (44) 01235 400414, Fax: (44) 01235 400454. Lines are open 9.00–6.00, Monday to Saturday, with a 24 hour message answering service. Email address: orders@bookpoint.co.uk

For USA & Canada orders: please contact NTC/Contemporary Publishing, 4255 West Touhy Avenue, Lincolnwood, Illinois 60646–1975, USA. Telephone: (847) 679 5500, Fax: (847) 679 2494.

Long renowned as the authoritative source for self-guided learning – with more than 30 million copies sold worldwide – the *Teach Yourself* series includes over 200 titles in the fields of languages, crafts, hobbies, business and education.

A catalogue record for this title is available from The British Library

Library of Congress Catalog Card Number: On file

First published in UK 1988 by Hodder Headline Plc, 338 Euston Road, London, NW1 3BH.

First published in US 1993 by NTC/Contemporary Publishing, 4255 West Touhy Avenue, Lincolnwood (Chicago), Illinois 60646–1975 USA.

The 'Teach Yourself' name and logo are registered trade marks of Hodder & Stoughton Ltd.

Copyright © 1999 Alan Graham

In UK: All rights reserved. No part of this publication may be reproduced or transmitted in any form or by any means, electronic or mechanical, including photocopy, recording, or any information storage and retrieval system, without permission in writing from the publisher or under licence from the Copyright Licensing Agency Limited. Further details of such licences (for reprographic reproduction) may be obtained from the Copyright Licensing Agency Limited, of 90 Tottenham Court Road, London W1P 9HE.

In US: All rights reserved. No part of this book may be reproduced, stored in a retrieval system, or transmitted in any form, or by any means, electronic, mechanical, photocopying, or otherwise, without prior permission of NTC-Contemporary Publishing Company.

Cover from Jacey/Début Art
Typeset by Wearset, Boldon, Tyne and Wear
Printed in Great Britain for Hodder & Stoughton Educational, a division of Hodder Headline Plc, 338 Euston Road, London NW1 3BH by Cox & Wyman Ltd, Reading, Berkshire.

Impression number	10	9	8	7	6	5	4		
Year		2005	2004	2003	2002	2001			

CONTENTS

ABOUT THE AUTHOR

Alan Graham has been a lecturer in Mathematics Education at the Open University since 1977. His particular interest is statistics and he has written a number of books and OU course units in this field. He has given numerous workshops and lectures to teachers on a variety of themes, including calculators, spreadsheets, practical mathematics and statistics.

ACKNOWLEDGEMENTS

Many thanks to Irene Dale, Tracy Johns, Ian Litton and Kevin McConway for their help and support.

FOREWORD

Most statistics textbooks are too hard. This is not the fault of the people who read them but is due to the unrealistic expectations of the people who write them. Most learners starting statistics are, unsurprisingly, not very confident with handling numbers or formulas, so why pretend they should be?

This book is my attempt to present the basic ideas of statistics clearly and simply. It does not provide a comprehensive guide to statistical techniques – if you want that, you will need a different sort of book. Instead, the focus is on understanding the key concepts and principles of the subject. Increasingly, much of the technical side of statistical work – performing laborious calculations, applying complicated formulas and looking up values in statistical tables – is these days handled on computers and calculators. What really matters is to understand the principles of what you are doing and be able to spot other people's statistical jiggery pokery when you come across it.

I believe that the ideas in this book will provide you with a bedrock understanding of what statistics is all about. I have enjoyed writing it. I hope that you will enjoy reading it.

Alan Graham 1999

1 | INTRODUCING STATISTICS

Some good reasons to learn about statistics

Statistics are all around us – in newspapers, on television, in general conversation with friends. Here are a few typical examples of headlines, graphs and advertisements that you might easily find in a daily newspaper.

Although it may not yet be immediately obvious to you, each of these examples embodies an important idea in statistics. They suggest just a few of the choices that we make all the time, with or without an understanding of statistical ideas. An underlying aim of this book is to suggest to you that statistics isn't just something you learn about from a textbook, but it can actually change the way you see the world and inform your judgements. There are a number of more specific reasons why it makes sense to have a basic grasp of statistics. Below are listed just some of them.

- As these examples suggest, much of the information we have to process in our different life roles – at home, as consumers, at work, in the community and as citizens in a wider economic and political setting – comes in the form of numbers, graphs and charts. A statistical awareness helps us to understand the forces acting on us and to process the information that confronts us: particularly when it is being used in a deliberately misleading way.

- Many of the important decisions which we have to make involve handling numbers and weighing up risks. A knowledge of statistics and probability won't guarantee that you always make the *correct* decision, but at least your decisions should be better informed.

- From archaeology to zoology, almost every subject covered at school, college or in the media has become increasingly quantitative, so it is becoming ever harder to hide from the need to have a basic understanding of statistics. This is particularly true of business.

- Statistics can be enjoyable – see how you feel about this idea by the time you have finished this book!

What sort of statistical questions can be asked?

There seem to be three basic types of question that crop up regularly in statistics. These are listed below, with an example of each:

(a) Can you summarize the data?

Looking at a lot of facts and figures does not always provide you with a clear picture of what is going on. It is often a good idea to find a way of summarizing the information – perhaps by reducing the figures to just one representative figure. For example, in order to get a sense of the extent to which women smokers reduce their smoking during pregnancy, 1000 women smokers might be asked to record the number of cigarettes they smoked before and during pregnancy. The raw data will yield two sets of one thousand numbers each – too much information to take in at a glance. An obviously useful way of summarizing the data in this case would be to calculate the *average* number of cigarettes smoked before and during pregnancy.

There are a number of statistical techniques for summarizing data – for example, 'summary statistics' such as graphs, and averages. You can read about these in Chapters 2, 3, 4 and 5.

(b) Is there a significant difference between these two sets of results?

Many of the decisions we make are based on our judgement of whether one thing performs better, lasts longer, or offers better value for money than another. It is easy enough to show that one set of results is different from another, but just how significant that difference might be, is less easy to decide on. So, following on from the previous example, the researchers may find that they produce summary results similar to the following:

Average number of cigarettes smoked before pregnancy 15.2
during pregnancy 14.9

Clearly there *is* a difference between these two results (although it would be very surprising if they had turned out to be identical) but it is by no means obvious whether we can place any great *significance* on the degree of difference.

This is where the idea of 'significance tests' comes in. These give a way of assessing the differences that you observe to help you decide whether or not they are 'significantly different' from each other. This idea is discussed in Chapters 13 and 14.

(c) Is there a close relationship between the two things under consideration?

Sometimes we are less interested in differences between two sets of results than in exploring the possible relationship between two things. For example, the researchers on smoking may wish to try to identify the main factor, or factors, that are linked to smoking. They may use their data to test whether there was a close connection between, say, a person's level of *psychological stress* and the number of cigarettes smoked, or if *disposable income* provided a stronger link with smoking. (Of course, the factors of stress and income are themselves closely intertwined and part of the task for the statistician is to try to isolate each of the factors that they choose to consider.)

This idea of a relationship between two things is discussed in Chapters 10 and 11. Chapter 10 is concerned with describing the sort of relationship, while Chapter 11 deals with measuring how strong the relationship is.

How are statistical questions investigated?

Data are usually collected in response to a question. The decision as to which technique is to be used will depend on the sort of question that has been asked at the start of the investigation. Posing a clear question is normally the first stage of any statistical work. For example, a doctor might be interested to investigate whether a new treatment for asthmatics is more effective than the old one. A sales manager may wish to find out which of a variety of different company advertisements was actually the most successful. A teacher could plan to catalogue and summarize the most common errors committed by students in a particular topic. These are the sorts of questions which give a purpose and a direction to statistical work and they breathe life and interest into the subject.

In fact, there are four clearly identifiable stages in most statistical investigations, which can be summarized as follows:

The stages of a statistical investigation

- Stage 1   pose a question
- Stage 2   collect relevant data
- Stage 3   analyse the data
- Stage 4   interpret the results

Not surprisingly, there are different statistical techniques suitable for the different stages. For example, issues to do with choosing samples and designing questionnaires will tend to crop up at the second stage, 'collect relevant data'. The third stage of analysing the data will probably involve various calculations (perhaps finding the average or working out a percentage) while the final stage of interpreting the results will raise ques-

	STAGES OF AN INVESTIGATION	TOOLS AND TECHNIQUES OF STATISTICS
1	Pose a question	
2	Collect relevant data	• choose a sample • design a questionnaire • conduct an experiment
3	Analyse the data	• calculate a percentage • calculate a mean • draw a helpful graph
4	Interpret the results	• make a prediction • test for cause-and-effect

Table 1.1   Connecting up the stages and the tools of statistics

tions such as whether the relationship under consideration is likely to be a cause-and-effect one. Table 1.1 summarizes some of these connections.

Although this book has been written with these stages clearly in mind, the chapters do not follow rigidly through them in sequence. For a start, there are no chapters specifically concerned with Stage 1, 'pose a question'. Also, it seemed sensible to devote the first part of the book (Chapters 1 to 6) to providing a basic explanation of some of the simpler statistical graphs and calculations. For example, Chapter 2 explains some basic ideas of maths that you will draw on later in the book. Thereafter, the second stage of 'collecting the data' is dealt with in Chapters 7 and 8 (on sampling and sources of data). Stage 3, 'analysing the data' deals with the tools and techniques of the subject and is central to Chapters 3 and 4 (on graphs), Chapter 5 (summary statistics) and Chapter 9 (reading tables of data). Lastly, the interpretation phase of statistical work is discussed in Chapter 6 (lies and statistics) and in the final five chapters of the book (covering regression, correlation, probability and statistical tests of significance). This is summarized in Table 1.2.

For a more comprehensive guide to the key ideas of statistics and how they are connected, turn now to the appendix at the back of the book. Here you will find a flow diagram which takes you, in a sensible sequence, through the sort of questions that users of statistics tend to ask. The diagram indicates the common statistical techniques that are required at each stage and where they can be found in this book. If at any time, when reading later chapters of the book, you are uncertain where the

THEME	CHAPTER
General introduction	1 Introducing statistics 2 Some basic maths
Pose a question	
Collect relevant data	7 Choosing a sample 8 Collecting information
Analyse the data	3 Graphing data 4 Choosing a suitable graph 5 Summarizing data 9 Reading tables of data
Interpret the results	6 Lies and statistics 10 Regression . . . 11 Correlation . . . 12 Introducing probability 13 Probability models 14 Testing for a difference

Table 1.2 Themes and chapters of the book

ideas fit into a wider picture, you may find it helpful to turn to this diagram again.

Choosing and using your calculator

Finally, you may have been wondering whether it would be a good idea to use a calculator (or perhaps buy a new one) to support your learning of statistics. Take courage! The calculator has a very positive role to play. In the past, a feature of studying statistics was the time students had to spend performing seemingly endless calculations. Fortunately, those days have gone, for, with the support of even a basic four-function calculator, the burden of hours spent on arithmetic is no longer necessary. These simple calculators are both cheap and reliable. However, if you are prepared to spend just a little more and buy a rather more sophisticated calculator, with customized statistical keys, the proportion of your time spent actually focusing on the central ideas of statistics will greatly increase. You may be thinking of buying a calculator, in which case, the keys which you will find particularly useful in statistical work are shown opposite. Don't worry if they seem unfamiliar – they will all be explained in later chapters.

$\bar{x}$ (x bar) calculates the average, i.e. the mean (see Chapters 2 and 5)

Σx (sigma x) adds all the x values together (see Chapter 2)

Σx^2 (sigma x squared) adds the squares of the x values

n finds the number of values already entered

$x\sigma_n$ (x sigma n), sometimes written as σ_n, finds the standard deviation of the x values (see Chapters 2 and 5)

r finds the coefficient of correlation (see Chapter 11)

a, b finds the linear regression coefficients (see Chapter 10)

It is possible that some extra statistical keys are available on your calculator, but the chances are that these are just variants of the ones listed above.

Although this book has been written with such a calculator in mind, explanations will also be included based on pencil and paper methods only.

Finally, you may have access to a computer with a spreadsheet package. If so, this will be a most valuable tool for analysing data and gaining a better grasp of statistical ideas. Don't worry if you have never used a spreadsheet before – in Chapter 9 you will be provided with a basic introduction, from scratch, of how to use one.

The next chapter deals with a few important topics in maths which should help you to make the most of some of the later chapters. These are simple algebra, graphs and some of the mathematical notations commonly found in statistics textbooks.

2 | SOME BASIC MATHS

If you flick through the rest of this book, you will find that complex mathematical formulas and notations have been kept to a minimum. However, it is hard to avoid such things entirely in a book on statistics, and if your maths is a bit rusty you should find it worth spending some time working through this chapter. The topics covered are algebra, simple graphs and statistical notation and they offer a helpful mathematical foundation, particularly for Chapters 5, 10 and 11.

Algebra

Algebraic symbols and notation crop up quite a lot in statistics. For example, statistical formulas will usually be expressed in symbols. Also, in Chapter 10 when you come to find the 'best-fit' line through a set of points on a graph, it is usual to express this line algebraically, as an equation.

Clearly it would be neither possible nor sensible to attempt to cover all the main principles and conventions of algebra in this short section. Instead, your attention is drawn to two important aspects of algebra which are sometimes overlooked in textbooks. The first of these is very basic – *why do we use letters at all* in algebra and what do they represent? The second issue concerns *algebraic logic* – the agreed conventions about the order in which we should perform the operations of $+$, $-$, $\times$ and $\div$.

Why letters?

Can you remember how to convert temperatures from degrees Celsius (formerly known as Centigrade) to degrees Fahrenheit? If you were able to remember, the chances are that your explanation would go something like this:

Suppose the original temperature was, say, 20°C. First you multiply this by 1.8, giving 36 and then you add 32, so the answer is 68.

The explanation above, which is perfectly valid, is based on the common principle of using a *particular* number, in this case a temperature of 20°, to illustrate the *general* method of doing the calculation. It isn't difficult to alter the instructions in order to do the calculation for a different number, say 25 – simply replace the 20 with 25 and continue as before. What we really have here is a formula, in words, connecting Fahrenheit and Celsius. A neater way of expressing it would be to use letters instead of words, as follows:

$$F = 1.8C + 32$$

(where F = temperature in degrees Fahrenheit and C = temperature in degrees Celsius).

Reading the formula aloud from left to right, you would say 'F is equal to one point eight C plus thirty two'.

The formula is equivalent to the word description shown earlier but here the letter C stands for whatever number you wish to convert from °C to °F. The F is the answer you get expressed in °F. The choice of letters for a formula is quite arbitrary – X and Y are particular favourites – but it makes sense to choose letters so that it is easy to remember what they stand for (hence F for Fahrenheit and C for Celsius in the formula shown above). The main point to be made here is that, in algebra, each letter is a sort of place-holder for whatever number you may wish to replace it with. A formula such as the one above provides a neat summary of the relationship that you are interested in and shows the main features at a glance.

There are certain conventional rules in algebra which you need to be clear about. Firstly, have a look at the following examples where a number and a letter are written close together:

$$5x; \; 1.8C; \; -3y$$

Although it doesn't actually say so, in each case the two things, the number and the letter, are to be *multiplied* together. Thus, $5x$ really means 5 times x. Similarly, $1.8C$ means 1.8 times C, and so on.

Let us now go back to the temperature conversion formula. The table on page 10 breaks it down into its basic stages, with an explanation added at each stage.

	MEANING
$F =$	The temperature, in degrees F, is equal to ...
$1.8C$	... 1.8 times the temperature in degrees C ...
$+32$	... plus 32

Table 2.1 The temperature formula, $F = 1.8C + 32$, stage by stage

---

EXERCISE 2.1. Using the temperature formula

Use the formula $F = 1.8C + 32$ to convert the following temperatures into degrees F.

   a) $0°C$
   b) $100°C$
   c) $30°C$
   d) $-17.8°C$

Comments on page 23

A formula is really another word for an *equation*. Usually a formula is written so that it has a single letter on the left of the equals sign (in this case the F) and an expression containing various numbers and letters on the right. We say that the equation $F = 1.8C + 32$ is *a formula for F* (F is the letter on the left of the equals sign) *in terms of C* (C is the only letter, in this case, on the right hand side of the equals sign).

To *satisfy* a formula or an equation is to find values to replace the letters which will make the two sides of the equation equal. For example, the values $C = 0$, $F = 32$ will satisfy the equation above. We can check this by *substituting* the two values into the equation, thus:

$$F = 1.8C + 32$$
$$32 = 1.8 \times 0 + 32 \ldots \text{which is true!}$$

If you feel you need a little practice at this idea, try Exercise 2.2 now.

Algebraic logic

A second important convention in algebra is the *sequence* in which you are expected to do calculations within a formula. In the previous example, notice that the temperature in °C is multiplied by 1.8 *before*

EXERCISE 2.2. Satisfying equations

Which of these pairs of values satisfy the equation $Y = 3X - 5$?

    a) $X = 3, Y = 4$
    b) $X = 0, Y = 5$
    c) $X = -2, Y = -11$

Comments on page 23

you add the 32. In general, where you appear to have a choice of whether to add, subtract, multiply or divide first, the multiplication and division take precedence over addition and subtraction. This would be true even if the formula had been written the other way round, like this:

$$F = 32 + 1.8C$$

This may seem odd but the reason for the convention is simply that, otherwise, the formula would be ambiguous. So it doesn't matter which way round the formula is written; provided the multiplication is done first, the result is still the same. Many calculators are programmed to obey this convention of algebraic precedence. If you have a calculator to hand, this would be a good opportunity to check out its logic, if you haven't already done so.

EXERCISE 2.3. Guess and press

Guess what result you think you would get for each of the following key sequences and then use your calculator to check each guess.

KEY SEQUENCE	YOUR GUESS	PRESS AND CHECK
2 ⊠ 3 ⊟ 4 ⊟		
2 ⊞ 3 ⊠ 4 ⊟		
2 ⊟ 3 ⊠ 4 ⊟		
2 ⊟ 3 ⊡ 4 ⊟		

Comments on page 24

As a result of the previous exercise, you should now be clear about the nature of the 'logic' on which your calculator has been designed. However, whether your calculator is programmed with 'arithmetic' logic, or 'algebraic' logic, written algebra obeys the rules of algebraic logic. We can take as an example, the second key sequence shown in Exercise 2.3. If you were to read this sequence to a human calculator, the chances are that they would produce the answer 20 ($2 + 3 = 5$, times $4 = 20$). However, suppose the question is now asked in the following slightly different way:

Find the value of $2 + 3X$, when $X = 4$.

This still produces the same calculation ($2 + 3 \times 4$), but the rules of algebra require that the sequence of the operations gives higher priority to the multiplication than the addition. If the spacing is changed slightly, this idea is easier to grasp, thus:

$$2 \quad + \quad 3 \times 4$$

When the multiplication is done first, the algebraically 'correct' answer is 14.

There is not room here to develop this notion of the logic of algebraic precedence further (i.e. the priorities given to each operation), but you will find that a good understanding of the ideas described above is helpful if you wish to handle algebraic expressions and equations successfully.

You might be wondering what happens if you want a formula involving multiplication and addition but where you wish the addition to be done before the multiplication. This is where the use of brackets comes into algebra. An example of this is where the temperature conversion formula is rearranged so that it will convert from degrees Fahrenheit into degrees Celsius. Remember that if you reverse a formula, not only do you have to reverse the operations ($\times$ becomes $\div$ and $+$ becomes $-$) but you must also reverse the sequence of the operations. Exercise 2.4 will give you a chance to work this out for yourself.

There are certain problems if we write the new conversion formula as:

$$C = F - 32 \div 1.8$$

If this formula were to be entered directly into a calculator with algebraic logic, it would give the 'wrong' answer. The reason is that, because divi-

EXERCISE 2.4. Reversing a formula

The middle column of Table 2.2 gives the two stages in the formula for converting temperatures from °C to °F.

> a) *Fill in the blank third column to show the two stages (in the correct sequence) needed to convert temperatures from °F to °C.*

SEQUENCE	CONVERTING FROM °C TO °F	CONVERTING FROM °F TO °C
First	multiply by 1.8	
Second	add 32	

Table 2.2   Converting to °C

> b) *Use your answer to part (a) to write down a formula which would allow you to convert temperatures from °F to °C.*

Comments on page 24 and continued below.

sion has a higher level of precedence than subtraction, a calculator with algebraic logic would divide 32 by the 1.8 before doing the subtraction. Putting in brackets, as shown below, takes away the ambiguity and forces the calculator to perform the calculation in the required sequence.

$$C = (F - 32) \div 1.8$$

To end this section, here are a few of the common algebraic definitions that you may need to be familiar with.

- *Expression:* $3X^2 - 4X + 5XY + 3X$ is an example of an expression.
- *Term:* There are four terms in the previous expression. They are, respectively, $3X^2$, $-4X$, $5XY$ and $3X$.
- *Sign:* The sign of a term is whether it is positive or negative. Only the second of these four terms is negative; the others are positive. When we are writing down a positive term on its own we don't normally bother to write in the

'+' sign before it. Thus we would write $5XY$, rather than $+5XY$.

- *Term type:* This refers only to the part of the term that is written in letters. Thus, the first term in the expression above is an 'X-squared' term, the second is an 'X' term, and so on.

- *Coefficients:* The coefficient of a term is the number at the front of it. Thus, the coefficient of the first X term is -4. The coefficient of the XY term is 5, and so on. The coefficient tells you how many of each term type there are.

- *Like term:* Two terms are said to be 'like terms' when they are of the same term type. Thus, since the second and fourth terms in the expression above are both 'X terms', they are therefore 'like terms'. The phrase 'collecting like terms' describes the process of putting like terms together into a single term. For example the second and fourth terms, $-4X + 3X$, can be put together simply as $-X$. This result arises from adding the two coefficients, -4 and 3, giving a combined coefficient of -1. We don't normally write this as $-1X$, but rather as $-X$.

- *Simplifying expressions:* This is the general term for collecting like terms. It is a simplification in that the total number of terms are reduced to just one each of every term type.

- *Expanding:* This normally refers to expanding out (i.e. multiplying out) brackets. Thus, the expression $5(2X + 3) - 3(6 - X)$ could be expanded out to give $10X + 15 - 18 + 3X$. The purpose of doing this is usually to enable further simplification. Thus, in this case we can collect together the two X terms and the two number terms to give the simplified answer $13X - 3$.

- *Equations:* Equations look rather like expressions except that they include an equals sign, '$=$'. In fact, an equation can be defined as two expressions with an equals sign between them. The expression to the left of the equals sign is, for obvious reasons, called the 'left hand side', or 'LHS', while the other expression to the right of the equals sign is the 'right hand side' or 'RHS'. Here is an example of an equation: $2X - 3 = 4 - 6X$. Often, however, equations are written so that the RHS expression is zero. For example, $3 - 4X + X^2 = 0$.

■ *Solving equations:* Equations usually exist in mathematics text-books to be 'solved'. Solving an equation means finding the value or values for X (or whatever the unknown letter happens to be) for which the equation is true. For example, the simple equation $3X - 5 = 7$ holds true for one value for X. The solution to this equation is $X = 4$. This can be checked by 'substituting' the value $X = 4$ back into the original equation to confirm that the LHS equals the RHS. Thus, we get LHS $= 3 \times 4 - 5$, which equals 7. This is indeed the value of the RHS. If you were to try any other value for X, say $X = 2$, the equation would not be 'satisfied'. Thus, LHS $= 3 \times 2 - 5$, which equals 1 not 7. The equation $3 - 4X + X^2 = 0$ holds true for two values of X, $X = 1$ and $X = 3$, so there are two solutions to this equation.

And that is all the algebra we will touch on here. The next section deals with graphs and how equations can be represented graphically.

Graphs

Perhaps more than any other mathematical topic, the basic idea of a graph keeps cropping up in statistics. The reason for this is to do with how the human brain processes information. For most people, scanning a collection of figures, or looking at a page of algebra doesn't reveal very much in the way of patterns. However, our brains seem to be much better equipped at seeing patterns when the information has been re-presented into a picture. The crucial difference is that in a picture (a graph or

---

EXERCISE 2.5. × marks the spot

Have a look at the cross, × in the line above. How would you explain to someone else exactly where the × is located on this page?

Comments follow

diagram, for example) each fact is recorded onto the page by its *position*, rather than by writing some peculiar, arbitrary squiggle (a number or a letter). In order to understand how graphs work, it is worth spending a few minutes thinking about the idea of position and what information you need to provide in order to define position exactly. Now do Exercise 2.5 on page 15.

Basically there are three pieces of information that you need to provide in order to fix exactly where the ✗ is located. These are:

- which (original) point on the page you started measuring from – known as the 'origin';
- how far from the origin the point is *across* the page – the *horizontal* distance; and
- how far from the origin the point is *up* the page – the *vertical* distance.

In mathematics, graphs are organized around these three principles. The 'page' is laid out in a special way so that the starting point (the origin) is clearly identified – normally in the bottom left hand corner. The two directions, horizontal and vertical, are normally drawn out as two straight lines and are known, respectively, as the *X* axis and the *Y* axis. Finally, in order to keep things as simple as possible, the two axes are provided with a scale of measure.

So, provided the page is laid out with the origin clearly identified, and the axes are suitably scaled, the point marked with an ✗ in Figure 2.1 can be described by giving just two numbers, in this case 10 and 20. These two numbers are called the *co-ordinates* of the point and they are usually written inside brackets, separated by a comma, like this:

$$(10, 20)$$

The first number inside the brackets, known as the *X* co-ordinate, measures how far you have to go along the *X* direction from the origin. The second number, the *Y* co-ordinate, gives the distance you go up the *Y* direction. Exercise 2.6 will give you practice at handling co-ordinates.

Having completed Exercise 2.6, it is clear just by looking at the pattern of points on the graph that there must be some special property in these co-ordinates which they all share. Have a look now at the co-ordinates and see if you can find any pattern in the numbers.

You may have spotted that, in each case, the *Y* co-ordinates are 10 more

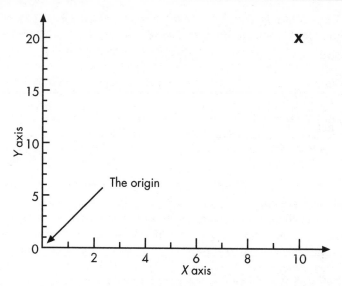

Figure 2.1 The axes of a graph

EXERCISE 2.6. Plotting co-ordinates

Plot the following co-ordinates on the graph in Figure 2.1.

(0, 10) (2, 12) (4, 14) (6, 16) (8, 18)

If you plot them correctly, the points should all lie on a straight line.

Comments on page 25

---

than their corresponding X co-ordinates. We can say this more neatly, using the language of algebra.

For each point, $Y = X + 10$

In fact, if a straight line were drawn through the points, this property ($Y = X + 10$) would apply to each and every of the many possible points that you might wish to choose that lay on the line. Each of the points that you plotted in Exercise 2.6 is said to 'satisfy' the equation. For example, the point (4, 14) consists of an X value ($X = 4$) and a Y value ($Y = 14$).

Substituting (4, 14) into the equation $Y = X + 10$
we get $\qquad\qquad\qquad\qquad\qquad 14 = 4 + 10$

To satisfy the equation means that this pair of values has produced a true statement. The full significance of this result is summarized below.

Graph   **Algebra**

| If a point lies on a line ... |

| .... the co-ordinates of the point 'satisfy' the equation of the line |

This is the crucial connection between graphs and algebra.

In the last example, you started with a set of points and deduced the equation of the line connecting them. In the next exercise, you will be asked to do this in reverse – i.e. you will be given a new equation, $Y = 2X - 3$, and asked to draw its graph by plotting various points.

EXERCISE 2.7. Drawing the graphs of equations

a) *Check that the point (2, 1) 'satisfies' the equation $Y = 2X - 3$.*

b) *Complete the co-ordinates below so that they satisfy the same equation. (0, ?) (1, ?) (?, 3) (4, ?)*

c) *Plot the point (2, 1) and the four points that you have worked out in (b) onto the graph, Figure 2.2. They should all lie on a straight line whose equation is $Y = 2X - 3$.*

Comments on page 24

You will have already noticed that, as with the previous example, the pattern of points appears as a straight line. It is worth commenting here that equations like $Y = X + 10$ and $Y = 2X - 3$ are known as '*linear*' equations for this reason – that when they are plotted onto a graph, they produce a *straight line*.

Finally, have a look at where the $Y = 2X - 3$ line passes through the Y axis. It is significant that it intersects the Y axis at the same value as the number term (-3) in the equation. This is a useful fact to remember about linear equations. The value -3 is known as the 'Y intercept'. The other number in the equation is the one associated with the X, in this case the number 2, and it will tell you how steep the line is – known as the

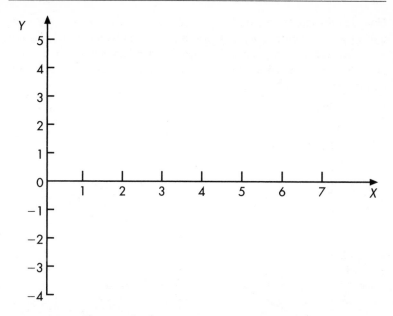

Figure 2.2   The graph of $Y = 2X - 3$

'slope' or 'gradient' of the line. It is a measure of how far the line goes up in the Y direction for every unit distance it goes across in the X direction. Expressed in more mathematical language, we can summarise all this by the following statement:

The general linear equation, $Y = a + bX$, has a *slope* of b and an *intercept* of a.

You may find it helpful to see a practical example of these ideas, so here is how it applies to working out your telephone bill. The explanation is shown below in three different forms: first in words, then in a formula and finally in the form of a graph.

How to calculate your quarterly telephone bill

(a)   in words
You pay a quarterly rental of £17.13 and then 4.4p per unit.

(b)   in a formula
$C = 17.13 + 0.044U$ (where C is the quarterly charge in pounds and U is the number of units used.)

Note that the charge rate of 4.4 pence per unit has had to be changed to 0.044 pounds, in order for the units to be consistent with the 17.13 figure, which is expressed in pounds. In general, when handling formulas, care needs to be taken to ensure that the units of measure are compatible with each other.

(c)   in a graph

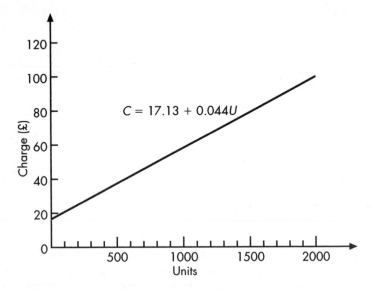

Figure 2.3   Graphing the telephone bill

This graph has produced the familiar 'linear' pattern. As you can see, the intercept of the graph gives the fixed part of the formula (the £17.13), whereas the slope is a measure of the charge rate per unit. Provided you know how many units you have used for a particular quarter, you can use the graph to read off the bill for that quarter. The next exercise asks you to do this now.

EXERCISE 2.8. Working out your bill from the graph

Suppose your quarterly consumption was 1000 units. From the graph, read off roughly what bill you would expect to have to pay.

Comments on page 26

Now to end this section on graphs, here are a few questions to get you thinking a bit more deeply about their properties.

EXERCISE 2.9. Some further questions about graphs

  a) *If $Y = a + bX$ is the general form of the linear equation, what does the general equation $Y = mX + c$ represent?*

  b) *A series of linear equations have all got the same intercepts but different slopes. Draw a sketch of what you think they will look like if they were plotted on the same axes.*

  c) *A series of linear equations have all got the same slopes but different intercepts. Draw a sketch of what you think they will look like if they were plotted on the same axes.*

Comments on page 26

Notation

There are one or two special symbols and notations in statistics that are in common use and are well worth familiarizing yourself with. If you have a specialized statistical calculator, you will already have run into some of these. The two described here are $\sum$ and $\bar{X}$.

(Capital) Sigma $\sum$

The first important symbol is the Greek letter capital sigma, written as $\sum$. In order to keep things interesting for you, statisticians have not one, but two symbols called 'sigma' in regular use. The lower case Greek letter sigma is written as 'σ' and it is normally used in statistics to refer to a measure of spread, called the 'standard deviation' (which is explained in Chapter 5).

This symbol, $\sum$, is an instruction to add a set of numbers together. So, $\sum X$ means 'add together all the X values'. Similarly, $\sum XY$ means 'add together all the XY products'. For example:

Table 2.3 shows all the X and Y values which you plotted as co-ordinates in Exercise 2.6.

X	Y
0	−3
1	−1
2	1
3	3
4	5
ΣX = 0 + 1 + 2 + 3 + 4 = 10	ΣY = (−3) + (−1) + 1 + 3 + 5 = 5

Table 2.3 Summing the X and Y values separately

To find ΣXY, it is necessary to calculate all the five separate products of X times Y and then add them together, thus:

X	Y	XY
0	−3	0
1	−1	−1
2	1	2
3	3	9
4	5	20
		ΣXY = 0 + (−1) + 2 + 9 + 20 = 30

Table 2.4 Summing the XY products

EXERCISE 2.10. Calculating with Σ

Use the data from the example above to calculate the following:

   a) ΣX^2
   b) ΣY^2
   c) $\Sigma (X + Y)$

Comments on page 27

X bar, $\bar{X}$

The symbol $\bar{X}$ is pronounced 'X bar' and refers to the mean of the X values. A 'mean' is the most common form of average, where you add up all the values and divide by the number of values you had. To take the previous example, the five X values are 0, 1, 2, 3, and 4. In this example, the mean value of X, is written as

$$\bar{X} = \frac{\sum X}{n}$$

Thus, $\bar{X} = \dfrac{0 + 1 + 2 + 3 + 4}{5} = \dfrac{10}{5} = 2$

Summary

This chapter introduced three basic mathematical ideas which underpin a lot of statistical work. These were algebra, graphs and notation. The section on algebra looked at two important aspects of this branch of mathematics – why we bother to use letters at all and the idea of algebraic logic (which controls the sequence in which algebraic operations are performed). The section on graphs explained how co-ordinates are plotted and introduced the terms *slope* and *intercept* of linear (i.e. straight line) equations. An important link was made between an algebraic equation and what it looks like when drawn as a graph. The final section explained two useful notations in statistics – the capital sigma ($\sum$) which means that a set of numbers are to be added together, and the use of a bar over a letter (for example, $\bar{X}$) to signify the *mean* of a set of values.

Comments on exercises

Exercise 2.1

    (a)   32°F (freezing point of water)

    (b)   212°F (boiling point of water)

    (c)   86°F

    (d)   −0.04°F (or roughly zero)

Exercise 2.2

We can substitute each of these pairs of values in turn into the equation to see if they give a correct result.

$$Y = 3X - 5$$

    (a) $X = 3, Y = 4$ $4 = 3 \times 3 - 5 \ldots$ which is true

   (b) $X = 0, Y = 5$ $5 = 3 \times 0 - 5 \dots$ which is not true

   (c) $X = -2, Y = -11 - 11 = 3 \times (-2) - 5 \dots$ which is true

Exercise 2.3

In terms of their inner 'logic', calculators come in two basic types. At the cheaper end of the market are the so-called 'arithmetic' calculators which perform calculations from left to right in the order in which they are fed in. The more sophisticated (and, generally, more expensive) 'algebraic' calculators are programmed to obey the conventions of algebraic precedence described earlier. Depending on which calculator you have, here are the results you are likely to get.

KEY SEQUENCE	'ARITHMETIC' ANSWER	'ALGEBRAIC' ANSWER
2 ⊠ 3 ⊟ 4 ⊟	2	2
2 ⊞ 3 ⊠ 4 ⊟	20	14
2 ⊟ 3 ⊠ 4 ⊟	−4	−10
2 ⊟ 3 ⊡ 4 ⊟	−0.25	1.25

Exercise 2.4

   (a)

SEQUENCE	CONVERTING TO °F	CONVERTING TO °C
First	multiply by 1.8	*subtract 32*
Second	add 32	*divide by 1.8*

   (b) A suitable formula would be: $C = (F - 32) \div 1.8$

Exercise 2.5

No comments

Exercise 2.6

The six points are located on the graph as shown on page 25.

Exercise 2.7

   (a) Substituting the values $X = 2$, $Y = 1$ into the equation $Y = 2X - 3$, we get:

         $1 = 2 \times 2 - 3 \dots$ which is true.

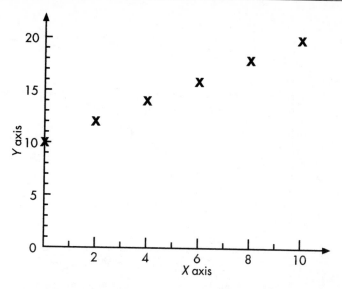

(b) The other four co-ordinates are $(0, -3)$, $(1, -1)$, $(3,3)$ and $(4,5)$

(c) The five co-ordinates and graph of the equation should look like this:

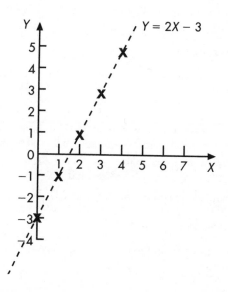

Exercise 2.8

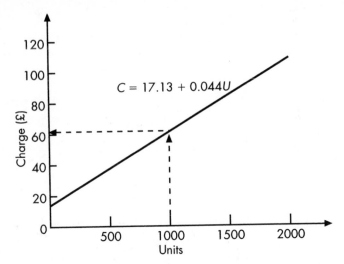

To make an estimate of your bill, follow the procedure shown on the graph above. The stages involved are described below.

    (a) start at 1000 units on the horizontal axis

    (b) draw a line vertically upwards until it meets the graph

    (c) from this point on the graph, draw a line horizontally across to the vertical axis

    (d) read off this axis the charge you will have to pay (about £61).

Exercise 2.9

    (a) These two equations may look different but they actually have the same form. In the second version (which is popular in some maths textbooks) the '*m*' is the letter used to represent the slope and the '*c*' is used for the intercept. The ordering of the two terms to the right hand side of the equals sign is quite arbitrary and of no mathematical significance.

    (b) These lines all have the same intercepts but different slopes.

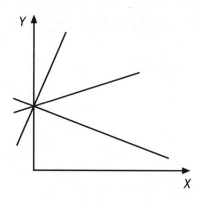

(c) These lines have all got the same slopes but different intercepts.

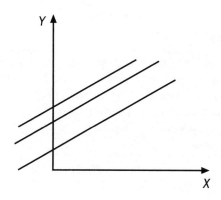

Exercise 2.10
The table below sets out the calculations that are required.

X	Y	X^2	Y^2	$X + Y$
0	−3	0	9	−3
1	−1	1	1	0
2	1	4	1	3
3	3	9	9	6
4	5	16	25	9
		$\Sigma X^2 = 30$	$\Sigma Y^2 = 45$	$\Sigma(X + Y) = 15$

3 GRAPHING DATA

This chapter introduces the main features of the following graphs – bar charts, pie charts, pictograms, histograms, stemplots, scattergraphs and time graphs. Boxplots are explained in Chapter 5. Related issues to do with *when* these graphs might be usefully drawn and *how they might be interpreted* are discussed in Chapter 4, so try not to get too interested at this stage in where the data came from and the patterns they may suggest to you!

Bar charts

The bar chart is probably the most familiar of all the graphs that you are likely to see in newspapers or magazines. Figure 3.1 is a typical example, which summarizes the popularity of different methods of contraception as reported to a particular family planning clinic.

Table 3.1 shows the data from which the bar chart has been drawn.

METHOD	PERCENTAGE (%)
Oral contraceptives	51
IUD	13
Cap/diaphragm	8
Sheath	15
Other	4
None	9

Table 3.1 Family planning clinic: methods of contraception

The chief virtue of a bar chart is that it can provide you with an instant picture of the relative sizes of the different categories. This enables you to see some of the main features from a glance at the bar chart – for example that oral contraceptives are by far the most popular method, that roughly twice as many use the sheath as the cap and so on. (Incidentally,

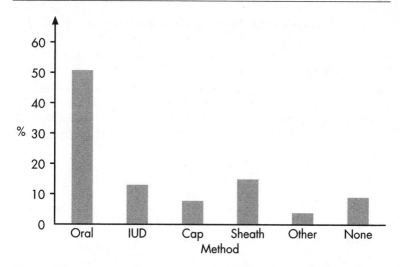

Figure 3.1   Family planning clinic: methods of contraception recommended or chosen
Source: estimated from UK national data

since these figures have been collected by a family planning clinic, they reflect only the preferences of those people who attended and it would be unwise to infer that they are typical of all family planning clinics or of the population as a whole.) There are a number of characteristics of a bar chart that are worth drawing attention to.

Firstly, you can see that the bars have all been drawn with the *same width*. This makes for a fair comparison between the bars in that the height of a particular bar is a measure of the frequency with which that category occurs.

Next, notice that it has been drawn with *gaps* between the adjacent bars. The reason for this is to emphasize the fact that the different methods of contraception listed along the horizontal axis are quite *separate categories*. This contrasts with a seemingly similar type of graph called a histogram, where the horizontal axis is marked with a continuous number scale and adjacent bars are made to touch. We will return again to this point later.

Now, have a look at the *order* in which the various methods of contraception have been placed along the horizontal axis of the bar chart. You may have wondered why this particular order has been chosen. The reason in this case is historical. The clinic in question first collected this sort of contraceptive preference data twenty years previously. The different methods have been ranked in order from most common to least common method for that particular year. However, over the intervening years the popularity of the IUD declined and was overtaken by the sheath. So, if current data only are being plotted, it would be more helpful to re-order the categories to reflect the current preferences, as follows:

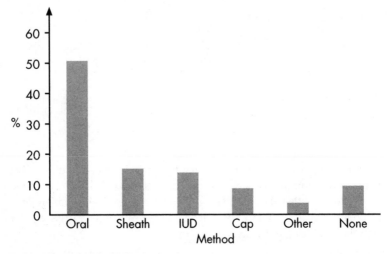

Figure 3.2   Reordered bar chart
Source: Table 3.1

Notice, however, that not all the bars have been placed in decreasing order of size. The final two categories, 'Other' and 'None', which seem to be distinct from the other 'methods' of contraception, have been placed on the right hand side of the bar chart simply to set them apart from the others.

The bars of a bar chart usually run vertically, as with Figure 3.2 above, but this need not always be so, as the next example shows.

If the full titles of the various contraceptive methods are to be displayed, there is simply not enough space to write in the names along the horizon-

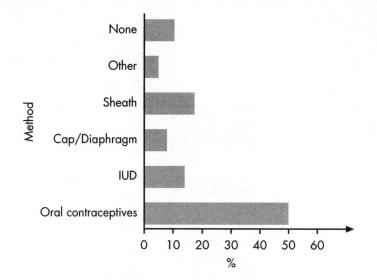

Figure 3.3 Horizontal bar chart showing the data from Table 3.1

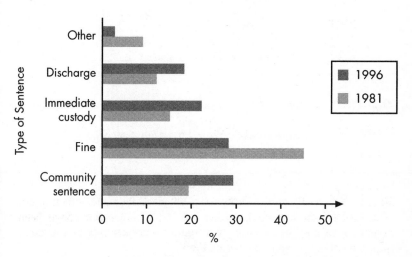

Figure 3.4 A compound horizontal bar chart showing the
offenders in England and Wales sentenced for
indictable offences 1981 and 1996
Source: *Social Trends 28*, 1998, Table 9.16

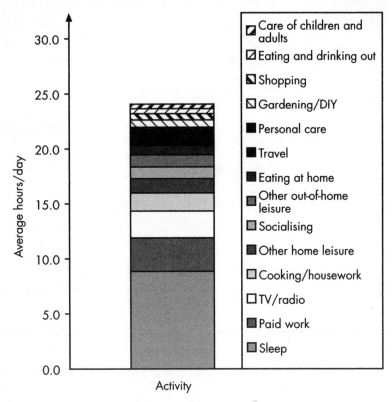

Figure 3.5 An unsatisfactory component bar chart showing use by economic activity
Source: Adapted from *Social Trends 28*, Table 13.2

tal axis. In this case, it makes sense to draw these bars horizontally rather than vertically, as shown in Figures 3.3 and 3.4. In general, then, horizontal bars are useful when there is quite a large number of categories and some of the category names are long.

A final subtlety worth mentioning is that bar charts are sometimes drawn so that the bars are grouped two or three at a time. This can result in two possibilities – the first is what is known as a *compound* or *multiple bar chart* as shown in Figure 3.4.

The box on the right of the compound bar chart that explains the shading on the bars is called the *legend*.

A bar chart can also be drawn with each bar split up to reveal its component parts. For obvious reasons, this is known as a *component bar chart* and a rather unsatisfactory example is shown in Figure 3.5.

Figure 3.5 was drawn to show how, on average, we spend our time in the course of a 24-hour day. As the title of this graph suggests, the component bar chart shown here is not satisfactory because there are too many components to allow a clear interpretation. Also, some of the categories are so small that the components tend to run into each other and so are rather difficult to distinguish. There are no hard and fast rules on the recommended number of components to use, but maximum of four or five seems sensible. A more helpful component bar chart is shown in Figure 3.6.

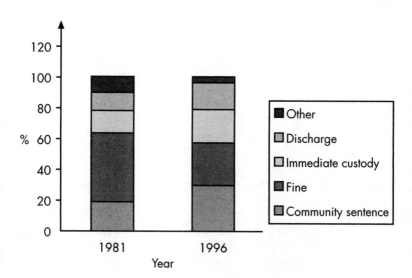

Figure 3.6   A component bar chart showing the offenders in England and Wales sentenced for indictable offences 1981 and 1996

Source: Figure 3.4

EXERCISE 3.1. Revision on bar charts

Here are a few questions to help you clarify for yourself some of the main points about bar charts.

a) *Why are bar charts drawn so that the bars are all the same width?*

b) *Why are bar charts drawn so that there are gaps between adjacent bars?*

c) *If there is no natural sequence to the categories on the horizontal axis of a bar chart, what would be a sensible basis on which to order the bars?*

d) *Distinguish between a compound bar chart and a component bar chart.*

Comments on page 45

Pie charts

Figure 3.7 shows the data from Table 3.1 represented in a pie chart.

As you can see from Figure 3.7, a pie chart represents the component parts that go to make up the complete 'pie' as a set of different-sized

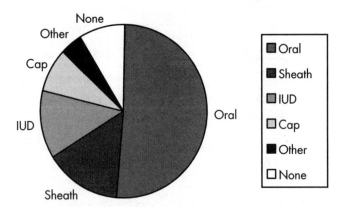

Figure 3.7   Pie chart showing methods of contraception recommended or chosen
Source: Table 3.1

slices. Drawing a pie chart by hand requires a little care, both in how the size of each slice is calculated and how it is drawn. The size of each slice as it appears on the pie chart will be determined by the angle it makes at the centre of the pie. You may remember that a full turn is 360° (360 degrees), so the angles at the centre in all the slices must add up to 360°. Taking Figure 3.7 as an example, the IUD slice should take up 13% of the pie, so the angle at the centre of that particular slice must be 13% of the entire 360°. If you have a calculator to hand calculate $360 \times \dfrac{13}{100}$ to get the answer 46.8°.

EXERCISE 3.2. Checking out the slices

> a) *Using either a calculator or pencil and paper, calculate the angles at the centre of each of the slices of the pie in Figure 3.7 and record your answers in Table 3.2.*
>
> b) *Either using a protractor or by eye, check that your answers below correspond to the angles in the slices drawn in Figure 3.7.*

METHOD	(%)	ANGLE (°)
Oral contraceptives	51	
IUD	13	$360 \times 13/100 = 46.8$
Cap/Diaphragm	8	
Sheath	15	
Other	4	
None	9	
Total	100	360

Table 3.2   Calculating the angles of the slices

Comments on page 45

Notice that the legend box has been included on the right of the pie chart in Figure 3.7. It isn't strictly necessary here, since the categories have also been written beside the corresponding slices of the pie. Normally either one or other of these approaches would be used as a means of identifying each slice of the pie.

Pictograms

Pictograms (also known as pictographs or ideographs) are really bar charts consisting of little pictures which indicate what is being represented. A simple example is shown in Figure 3.8.

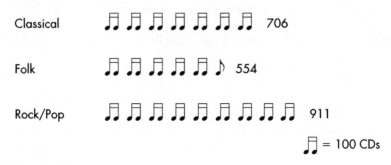

Figure 3.8 A pictogram showing the various numbers of CDs in a library, by type
Source: Personal survey

One or two aspects of the pictogram are worth clarifying. Firstly, notice the key at the bottom, which explains that one double-note symbol corresponds to 100 CDs. Also, you may have spotted that the final symbol in the Folk row has been 'cut in half' to indicate that the number of CDs in this category lies between 500 and 600. On final point: particularly where each line of the pictogram consists of several symbols, it is important to ensure that each symbol is the same size. The reason for this is to ensure that the length of each line of pictures is a consistent measure of the number of items it represents and that, visually, you are comparing like with like.

Histograms

The histogram in Figure 3.9 shows the proportions of babies born to mothers at different ages, in a typical UK hospital.

A key distinction between the histogram shown on page 37 and the bar chart described earlier is that here the adjacent bars are made to touch. The reason for this is to emphasize the fact that the horizontal axis on a histogram represents a continuous number scale. It is important to be aware that the bars on a histogram do not represent separate categories,

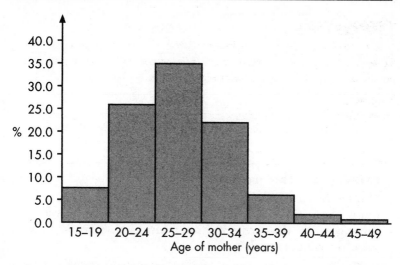

Figure 3.9 Histogram showing the number of live births by age of mother

Source: Estimated from national data

as they do on a bar chart, but, rather, adjacent intervals on a number line. In other words, although superficially the bar chart and the histogram look similar, they are designed to represent two quite different types of data. The bar chart is useful for depicting separate *categories*, while the histogram describes the 'shape' of data that have been *measured on a continuous number scale*. The distinction between these two types of data is an important one and will be returned to in the next chapter. In the example above, the data have been grouped into intervals of five years on the horizontal axis. This is known as the *class interval* and in Figure 3.9 the class intervals are all equal.

A further subtlety of a histogram which distinguishes it from the bar chart is that the widths of its bars (which correspond to the class intervals) need not all be the same. This may occur because the original data were collected into unequal class intervals in the first place or perhaps because you choose to collapse two or more adjacent intervals together. Suppose, for example, you wish to redraw Figure 3.9, collapsing together all the women aged over 30 years of age into one class interval. This would produce the *frequency table* shown in Table 3.3.

AGE OF MOTHER (YEARS)	FREQUENCY (%)
15–19	8.2
20–24	26.9
25–29	35.4
30–49	29.6

Table 3.3   Frequency table showing four adjacent class intervals collapsed together

EXERCISE 3.3. Spot the deliberate mistake

Figure 3.10 shows an *incorrect* histogram drawn from the data in Table 3.3. What do you think is wrong with the way it has been drawn and how would you correct it?

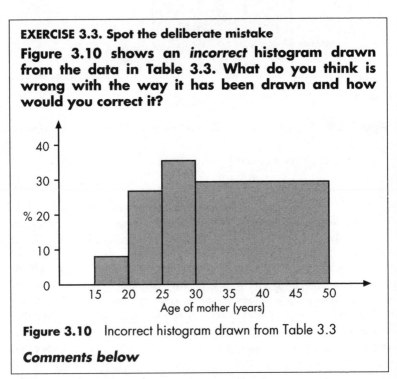

Figure 3.10   Incorrect histogram drawn from Table 3.3

Comments below

Comparing the overall shape of Figure 3.10 with that of Figure 3.9, it seems that the process of collapsing several class intervals together has the misleading effect of accentuating the importance of the wider interval. Since this interval is four times as wide as the other intervals, it makes sense to adjust for this by reducing the height of this bar to a quarter of the height shown in Figure 3.10, as in Figure 3.11.

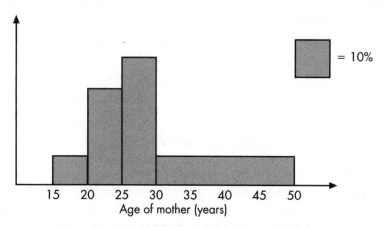

Figure 3.11 Corrected histogram drawn from Table 3.3

In effect, the height of the final wide column in Figure 3.11 is an average of the corresponding four separate bars in Figure 3.9 which it has replaced. What this means is that the total area for each of these histograms is the same – an important property of histograms and one which acts as a useful guiding principle if you ever wish to redraw them using different class intervals.

Finally, let's turn our attention to the scale on the vertical axis which shows the percentage of women contained in each age interval. Where all the class intervals are the same width, as is the case in Figure 3.9, it seems to be reasonably clear what these percentage figures refer to. However, for histograms like Figure 3.11, where the class intervals are unequal, a number scale on the vertical axis is less meaningful since the height of each bar must also be interpreted in terms of how wide it is. The key property which can be relied on is the *area* of each bar since, as was mentioned earlier, the total area of a histogram must be preserved no matter how the class intervals are altered. Thus, as is shown in Figure 3.11, it is more correct to draw a histogram without a vertical scale but, instead, with its unit area defined alongside the graph.

Stemplots

Stemplots, sometimes called 'stem-and-leaf' diagrams, can often be used as an alternative to histograms for representing numerical data. Table 3.4,

77	32	55	55	59	67	55	60	51	82
66	66	100	29	61	47	52	46	53	63
74	47	58	72	55	50	36	52	58	48
80	41	54	53	70	68	42	62	98	45

Table 3.4   The ages of 40 'famous' women listed in the Birthday column of a national newspaper

for example, gives some numerical data printed in a national newspaper over a period of several consecutive days. They have been taken from a Birthday column and reveal the ages of 40 women deemed to be sufficiently famous to warrant inclusion in the newspaper.

The stemplot started in Figure 3.12 shows how the first four ages are entered. As you can see, the vertical column on the left hand side, called the 'stem', represents the 'tens' digit of the numbers. The corresponding 'units' digits appear like leaves placed on the horizontal branches coming out from the stem. For example, the number 32 appears as a 2 (the units digit of the number) placed on level 3 (the tens digit of the number) of the stem, and so on.

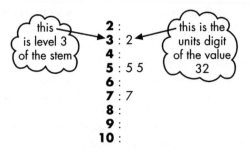

Figure 3.12   Incomplete stemplot representing the first four items from Table 3.4

EXERCISE 3.4. Completing the stemplot

Go through the remaining 36 ages in Table 3.4 and record their values in Figure 3.12.

Comments on page 46

The completed stemplot shown on page 46 has one important defect in that the leaves have been entered onto each stem in the order in which they appeared in Table 3.4. There is nothing special about this ordering and it makes sense to *sort* the stemplot so that the 'leaves' on each level are ranked from smallest to largest, reading from left to right. The final version, called a 'sorted' stemplot, is shown in Figure 3.13.

2 : 9
3 : 26
4 : 1256778
5 : 01223345555889
6 : 01236678
7 : 0247
8 : 02
9 : 8
10 : 0

10 : 0 is 100 years

Figure 3.13 'Sorted' stemplot based on the data from Table 3.4

As you can see, the stemplot looks a bit like a histogram on its side[1]. However, it has certain advantages over the histogram, the main one being that the actual values of the raw data from which it has been drawn have been preserved. Not only that but, when the stemplot is drawn in its sorted form, the data are displayed in rank order. These and other useful features of a stemplot will be looked at in more detail in the following chapter.

When two sets of data are to be compared, they can be drawn on two separate stemplots and placed back-to-back. Figure 3.14 shows a back-to-back stemplot to compare the age profiles of successful men and successful women, as represented by their inclusion in the 'birthday' columns of the same national newspaper.

There are interesting patterns evident in this particular back-to-back stemplot, but we will defer speculating about them until the next chapter.

Before leaving the stemplot, it is worth noting that the example chosen so

[1] In Figure 3.13 the stem has been drawn so that the stem values increase as you read downwards. Some textbooks show them the other way up. The main advantage of drawing them as shown here seems to be that, if rotated through 90°, the stemplot immediately looks like a histogram.

MEN   WOMEN

	2	9
8	**3**	26
99863	**4**	1256778
8633111110	**5**	01223345555889
997666442	**6**	01236678
9764411	**7**	0247
72211000	**8**	02
	9	8
	10	0

10 0 is 100 years

Figure 3.14   Back-to-back stemplot comparing 'successful' women's and men's ages

far to illustrate its use has used data which are two-digit numbers. With values like these, it is fairly obvious that the stem should represent 'tens' and the leaves should correspond to 'units'. However, if the data to be displayed were all decimal numbers less than 1 or very large numbers bigger than, say, 1000, then the stem and units would need to be redefined accordingly. In doing so, you may need to round the data to two significant figures, as the stemplot cannot easily cope with displaying numbers containing three or more digits.

Scattergraphs

Scattergraphs (sometimes known as scatterplots or scatter diagrams) are useful for representing paired data in such a way that you can more easily investigate a possible relationship between the two things being measured. Paired data occur when two distinct characteristics are measured from each item in a sample of items. Table 3.5, for example, contains paired data relating to twelve countries of the European Union (EU) – the area (in 1000 km^2) of each country and also their estimated populations (in thousands) for the year 2000. Figure 3.15 shows the same data plotted as a scattergraph.

Scattergraphs are a particularly important form of representation in statistics as they provide a powerful visual impression of what sort of relationship exists between the two things under consideration. For example, in this instance, there seems to be a clear pattern in the distribution of the

COUNTRY	AREA (1000 km²)	ESTIMATED POPULATION FOR AD 2000 (million)
Belgium	30.5	9.7
Denmark	43.1	5.2
France	544.0	61.0
Germany	357.0	77.65
Greece	132.0	10.1
Ireland	68.9	3.5
Italy	301.3	57.6
Luxembourg	2.6	0.4
Netherlands	41.2	15.7
Portugal	92.1	11.1
Spain	504.8	38.7
United Kingdom	244.1	58.8

Table 3.5   Area and estimated population of twelve EU countries
Source: Eurostat – basic statistics of the community, 1989

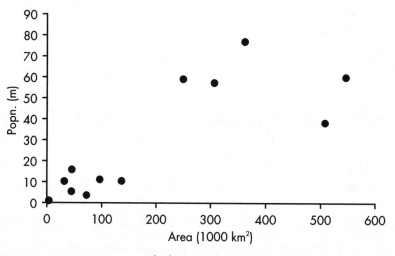

Figure 3.15   Scattergraph showing the relationship between population and area in twelve EU countries

points, which tends to confirm what you might have already expected from common sense, namely that countries with large areas tend also to have large populations. Understanding what is being revealed by a scattergraph is an important first step in getting to grips with two related statistical ideas which are examined in Chapters 10 and 11 – regression and correlation.

Time graphs

Figure 3.16 shows typical time graphs. Time is almost always measured along the horizontal axis and in this case the vertical axis shows the percentage of UK residents who took domestic holidays to four foreign locations over the period 1971 to 1996.

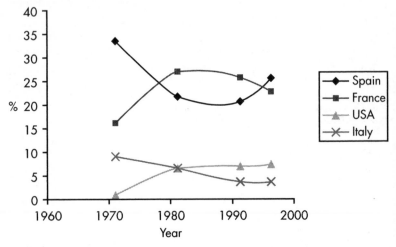

3.16 Line graphs showing holidays abroad by destination UK, 1971–1996
Source: *Social Trends 28*, 1998, Adapted from Table 13.22

There is an important difference between time graphs and scattergraphs. With time graphs, each point is recorded in regularly spaced intervals going from left to right along the time axis. Since they therefore form a definite sequence of points, it makes sense to join them together with a line, as has been done here. However, points plotted on a scattergraph are not consecutive and it would be incorrect to attempt to join the points on a scattergraph in this way.

Summary

This chapter introduced the main features of the following graphs – bar charts, pie charts, pictograms, histograms, stemplots, scattergraphs and time graphs.

Comments on exercises

Exercise 3.1

(a) Provided that the bars are the same width, the height of each bar gives a fair impression of its value. Where bars are unequal in width, your eye would tend to be drawn to the wider one and, intuitively, you might give it undue weighting.

(b) The gaps between the bars emphasize that each bar represents a separate category and there is not a continuous number scale on the horizontal axis.

(c) All things being equal, the most helpful ordering for the bars of a bar chart is usually to start with the tallest bar on the left and rank them in descending order from left to right.

(d) A compound bar chart shows a number of sets of bars grouped together on the same graph – usually in clusters of two or three at a time. With a component bar chart, each bar is drawn with its component parts shown stacked vertically within it.

Exercise 3.2

The table below shows the angles at the centre of each pie and how they were calculated.

METHOD	(%)	ANGLE (°)
Oral contraceptives	51	$360 \times 51/100 = 184$
IUD	13	$360 \times 13/100 = 47$
Cap/Diaphragm	8	$360 \times 8/100 = 29$
Sheath	15	$360 \times 15/100 = 54$
Other	4	$360 \times 4/100 = 14$
None	9	$360 \times 9/100 = 32$
TOTAL	100%	360°

Exercise 3.4

The complete stemplot looks like this:

```
 2 : 9
 3 : 26
 4 : 7678125
 5 : 55951238502843
 6 : 70661382
 7 : 7420
 8 : 20
 9 : 8
10 : 0
```

4 CHOOSING A SUITABLE GRAPH

In the previous chapter, a variety of different types of graph were merely described, without any real explanation of when and how they might be used. We now turn to this question of deciding which graph is most suitable for which situation.

Essentially, two key factors help you to determine which is the best graph to draw. These are the *type of data* being represented and the *type of statistical judgement* which you hope the graph will help you to make. These two aspects are discussed in the first two sections below. In the final section of the chapter, examples will be provided to give you practice at choosing the best graph to represent your data, guided by the principles of the first two sections.

Types of data

Statisticians have come up with a variety of sophisticated and elegant ways of classifying data according to a variety of different attributes. However, since the whole point of this section is to help you decide which graph to choose, we will restrict attention to a simple way of classifying data which informs the choice of graph.

Discrete and continuous data

The most basic distinction that can be made between types of data is to separate those which are *discrete* from those which are *continuous*. A dictionary definition of the word discrete will read something like 'separate, detached from others, individually distinct, discontinuous'. Table 4.1 shows a few examples of discrete data.

In all of the four examples shown in Table 4.1, each separate item of data collected can be placed into one of a limited number of predefined categories. For example, in the wild flower survey, if the next flower you find is a marigold, then it falls clearly into a 'separate, detached from

INVESTIGATION	TYPICAL ITEMS OF DATA
Types of wild flower	bluebell, marigold, meadow-sweet, …
Household size	1 person, 2 people, 4 people, 3 people, …
Environmental problems	oil spills, acid rain, dog fouling,…
Vehicle colour	red, green, white, …

Table 4.1   Examples of discrete data

others, individually distinct' category along with all the other marigolds. Similarly, in the household survey, there are a restricted number of separate household sizes – 1, 2, 3 people, and so on – into which the data can be recorded. Household sizes in between these values, say 2.73 people, or 3.152 people, for example, are simply not possible.

As you can see from these examples, discrete data can either take the form of *word categories* (bluebell, acid rain, etc) or *numbers* (the number of people in a household, for example). However, although word categories are always discrete, this is not always true of data in the form of numbers. Table 4.2 shows examples of numerical data which are not discrete and these form our second fundamental classification – known as *continuous data*.

INVESTIGATION	TYPICAL ITEMS OF DATA
Babies' birth weight	3120 g, 3760 g, 2700 g, …
Temperature survey	18.6°C, 21.4°C, 19.0°C, …
Survey of commuting times	23 mins, 11 mins, 70 mins, …

Table 4.2   Examples of continuous data

As you can see from these examples, continuous data are all numerical. However, unlike the survey of household size examples earlier, they are not restricted in the number of different values that they can take. Essentially the only restriction here lies in the degree of accuracy with which the measurements were taken. For example, a survey in which babies' birth weights are measured accurate to the nearest gram, will draw on ten times as many possible separate weights as a survey which weighed them accurate to the nearest ten grams. But the point is that, in theory, there is no limit to the number of possible weights at which a baby might be measured and this is a crucial property of continuous data in general.

Two further terms which are often used in this context are *attributes* and *variables*. An attribute is another term for a 'word category' and attributes will generate data that are non-numerical. For example, colour is an attribute and a survey of vehicle colours will product categorical data – red, green, blue, and so on. A variable, on the other hand, is a measure which generates numerical data. Thus, household size and babies' birth weight are both variables, the first producing discrete numerical data (1, 2, 3, etc) and the second producing continuous numerical data.

Confused? Well, now might be a good time to review what you have read so far.

EXERCISE 4.1. Consolidation

Have a look now at Table 4.3, and tick in the appropriate boxes to summarize the distinction between discrete and continuous data, variables and attributes.

	WORD CATEGORIES	DISCRETE NUMBERS	CONTINUOUS NUMBERS
Discrete data			
Continuous data			
Variables			
Attributes			

Table 4.3   Discrete and continuous, variables and attributes

Comments on page 63

It is quite helpful to have, in your mind, an image of continuous data as an infinite range of possible positions on a continuous number scale. Discrete data, on the other hand, whether formed from word categories or discrete numbers, can be thought of as a restricted collection of boxes into which each new item in the sample can be classified. Thus:

Word categories

Survey of types of wild flowers

| bluebells | | marigolds | | meadow-sweets |

Discrete numbers

Survey of household size | 2 people | | 3 people | | 4 people |

Continuous numbers

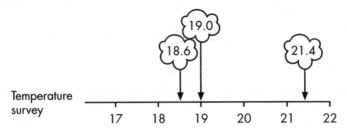

Later in the chapter a clear link will be made between the type of data you are dealing with (whether they are discrete or continuous) and the most suitable graph to draw, so it is important to be able to distinguish discrete from continuous measures. Exercise 4.2 will allow you to get some practice at this.

Although certain variables like length, weight, time and so on, can be described as 'continuous', truly continuous measurement is never attainable in practice. The reason for this is that there is no such thing as perfect accuracy in the real world. For example, if you need to measure a sparking plug gap or the thickness of a human hair, even with access to the world's most accurate ruler, you are eventually going to have to round your answer to a certain number of decimal places. So in a sense, all practical measurement is discrete.

Single and paired data

As was indicated in the section dealing with scattergraphs in the previous chapter, data can either be collected *singly* or *in pairs*. The meaning of these two terms is quite straightforward. If only one measure is taken from each item in the sample, the data have been collected singly. If two different characteristics are measured from each data item, then the data are said to be 'paired'. Here are a few examples of each.

Single data measurement

(a)	Weights of thirty eggs (nearest gram)	66, 73, 54, 59, 61 . . .
(b)	Temperature over thirty days (°C)	19, 17, 15, 18, 21, 14, . . .

EXERCISE 4.2. Discrete or continuous?

In the extract below, a number of items of data have been written in italics. Try to sort out which is discrete and which is continuous and record your answers in Table 4.4.

The dwarfing apple rootstock '*M27*' was raised in *1929* from a cross between 'M9' and 'M13'. As a dwarf bush, it makes a tree *1.2–1.5 m* in height and spread. A well-grown tree should yield on average *4.5–6.8 kg* of fruit each year. At planting, side-shoots are cut back to *three buds* and the leader pruned by about *one quarter*, cutting to an upward facing bud.

Adapted from Harry Baker, *Stepover Apples*, The Garden, March, 1991

DATA ITEM	TYPE OF DATA (D OR C)
M27	
1929	
1.2–1.5 m	
4.5–6.8 kg	
three buds	
one quarter	

Table 4.4   Classifying discrete and continuous data

Comments on page 63

(c) Type of wild flowers collected   bluebell,   marigold, meadow-sweet . . .

Paired data measurement

(a) Weights and sizes of thirty eggs (nearest gram)

EGG NUMBER	1	2	3	4	5	. . .
Weight (grams)	66	73	54	59	61	. . .
Size	2	1	5	4	3	. . .

(b) Average temperature (°C) and rainfall (cm) over twelve months

MONTH	JAN	FEB	MAR	APR	MAY	...
Temperature (°C)	7	9	11	12	16	...
Rainfall (cm)	17	14	11	9	8	...

Paired data are described more fully in Chapter 10.

Types of statistical judgement

Data are normally collected and graphed for a particular purpose – in order to help us make a statistical judgement. In Chapter 1, three of the most common types of statistical judgement that are made were described, and these are still fundamental to helping you decide what sort of graph you should draw. Firstly, graphs are often useful simply to *summarize* the data. Graphs are also useful for *comparing* and *inter-relating* and we will look at each of these three types of judgement in turn.

Summarizing

The most basic function of a graph is to summarize the essential characteristics of the data in question. You may be wondering, however, what exactly are the key characteristics that are worth examining; so let us start by having a look at the data in Figure 4.1, which have been taken from the previous chapter.

EXERCISE 4.3. The bare essentials

Jot down two or three of the essential features of these data that are revealed by the way they have been presented in this stemplot.

Comments below

Two key features are worth stressing here, both of which are ways of summarizing the data. The first is that a typical or average or 'central' value is around 55 years. This is a description of the *central tendency* of the data – a rather pompous statistician's way of describing where the data are centred. The second useful summary is to describe *how widely*

```
 2 : 9
 3 : 26
 4 : 1256778
 5 : 01223345555889
 6 : 01236678
 7 : 0247
 8 : 02
 9 : 8
10 : 0
```

10 : 0 is 100 years

Figure 4.1   Stemplot showing the ages of 40 'famous' women
Source: Figure 3.13

spread the data are. There are several ways of doing this, the easiest of which is just to state the lowest and the highest values. So, in this example, the range is 71 years, running from a low of 29 to a high of 100 years.

To recap, then, the two most common ways of summarizing data are firstly using a measure of central tendency (perhaps an average) and secondly by measuring the spread. Formal methods for performing these calculations are described in Chapter 5, but for the purposes of this chapter it is enough to be able to have an intuitive sense of them simply by looking at a graph.

Comparing

A fundamental question in statistics is whether one set of measurements is longer, heavier, warmer, longer-lasting, healthier, better value for money, etc than another. There are a variety of statistical tests designed to measure just how significant such differences are and some of these are explained in Chapter 14, *Testing for a difference*. However, before getting to grips with the notion of statistical tests of significance, it would be a good idea to develop an intuition about the idea of making such comparisons by purely graphical methods. Figure 4.2 shows a back-to-back stemplot comparing 'successful' women's and men's ages. (This was given in the previous chapter as Figure 3.14.)

MEN WOMEN

	2	9
8	**3**	26
99863	**4**	1256778
8633111110	**5**	01223345555889
997666442	**6**	01236678
9764411	**7**	0247
72211000	**8**	02
	9	8
	10	0

10 0 is 100 years

Figure 4.2 Back-to-back stemplot comparing 'successful' women's and men's ages
Source: Figure 3.14

EXERCISE 4.4. Interpreting a back-to-back stemplot

 a) *How does the back-to-back stemplot in Figure 4.2 help you to make a comparison between the two batches of data?*

 b) *What additional calculations might help you to make this comparison more fully?*

Comments below

(a) It is clearly evident from Figure 4.2 that the women's ages are more widely spread than those of the men (a range of 29–100 years, as opposed to 38–87 years for the men). Also, the women were, overall, slightly younger than the men. (Fifteen men were in the 'seventy and over' category, as opposed to only eight women.)

(b) While this back-to-back stemplot may give you a reasonably good intuitive feeling for the differences between the two batches of data, it would also be helpful in this case to calculate the averages of the women's and men's ages. The most common form of

average is the mean – found by adding all the values in the batch together and dividing by the number of values in the batch. The mean age of the men in Figure 4.2 turns out to be 64.2 years, while the mean age of the women is only 58.5 years, confirming the impression conveyed by the back-to-back stemplot that the men are rather older than the women. (The mean and other types of average are explained more fully in Chapter 5.)

Inter-relating

The third of our fundamental statistical judgements is inter-relating. This means finding a way of describing the relationship between two variables. For example, a manufacturer might be interested to see whether there is a link between sales and the amount of money spent on advertising. A scientist may wish to explore the connection between the temperature of a metal bar and how long it is. An important relationship in medicine is the connection between the drug dosage and its effects on patients. Environmentalists may want to investigate the link between the burning of fossil fuels and resulting changes in the weather. In each of these cases, there are not one but two separate variables being considered and it is their inter-relationship which is of particular interest. The most likely approach will be to collect and analyse paired data taken from each of the variables together. For example, Table 4.5 provides paired data from which you might be able to investigate a possible inter-relationship between someone's height at age 2 years and their eventual adult height.

NAME	HEIGHT AT AGE 2 YEARS (m)	ADULT HEIGHT (m)
Alan	0.84	1.71
Hilary	0.80	1.56
Luke	0.89	1.80
Ruth	0.72	1.50
Ian	0.86	1.76
Pauline	0.83	1.65
Bal	0.88	1.72
Marti	0.79	1.60
Francis	0.85	1.73
Iris	0.77	1.55

Table 4.5   Paired data showing height at age 2 years and eventual adult height
Source: Personal data

The most helpful type of graph for revealing inter-relating patterns in paired data is the scatterplot.

Getting the right graph

So far in this chapter you have been introduced to two key elements which inform your choice of graph. These are the *type of data* being represented and the *type of statistical judgement* which you hope the graph will help you to make. You may already have seen that these two elements are not entirely separate from each other since, not surprisingly, the sort of data that you decide to collect are bound to be related to the sort of decision you intend to make about them. A clear example of this is with paired data, which you would almost certainly have collected in order to explore an inter-relationship. In this final section some examples are provided which will tie these ideas together with what you have learnt about the key features of the common graphs from Chapter 3. You will be asked to look at some of the graphs in Chapter 3 and make an assessment of them in terms of the two key elements described above. In order to simplify the task, this box provides you with a handy summary of the key ideas introduced so far.

Type of data:	discrete	continuous	
Type of judgement:	summarizing	comparing	inter-relating

Bar charts and pie charts

Let us start with bar charts and pie charts.
Methods of contraception are categories so these are discrete data in both

---

EXERCISE 4.5. Analysing bar charts

Have a look at Figures 3.1 and 3.4 and for each of these bar charts use the summary above to identify

> a) *the type of data being represented*
> b) *the type of statistical judgement which they might enable you to make.*

Comments below

---

cases. However, the type of statistical judgements which they inform are rather different. Whereas the simple bar chart in Figure 3.1 merely *describes* the relative popularities of the various methods of contraception for a given year, the compound bar chart in Figure 3.4 is helpful for *comparing* two different batches of data, one for the year 1976 and the other for 1988. Now you may have felt that the simple bar chart did enable you to make comparisons – comparing between its various columns. However, this is not the sense in which the term comparing was originally used on page 54. Comparing, as the term is used here, is concerned with how one *set* of measurements differs from another.

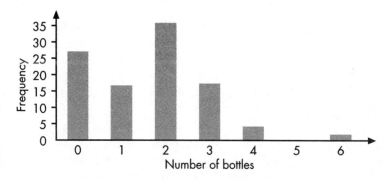

Figure 4.3   A bar chart used to represent discrete numerical data
– the number of milk bottles on 100 doorsteps
Source: Personal survey

In general, a bar chart should be used only when the quantity on its horizontal axis is discrete. This may be a set of categories drawn from an attribute, as in the case of Figure 3.1. Alternatively, the data may be discrete numerical, such as in the example in Figure 4.3.

If you wish to compare two batches of discrete data, then either use two separate bar charts, a compound bar chart or a component bar chart.

We now turn to pie charts.

EXERCISE 4.6. Analysing pie charts

Have a look at the pie chart in Figure 3.7 and try to identify:

   a) *the type of data being represented*
   b) *the type of statistical judgement which it might enable you to make.*

Comments below

Your responses here should have been the same as for the bar chart in Figure 3.1, namely that the data being represented are *discrete* and the judgement is one of *describing*. As before, a case could be made for claiming that the pie chart allows you to make comparisons between the slices of the pie. However, this is not the sense in which the term 'comparing' is being used here.

In general, like the bar chart, a pie chart should be used only to depict discrete data. These may be a set of categories drawn from an attribute, as in the case of Figure 3.1. However, unlike for the bar chart, it does not make sense to draw a pie chart for data which are discrete numerical.

EXERCISE 4.7. A rather half-baked pie chart?

Have a look at the example in Figure 4.4, and try to decide why this pie chart may not be the ideal representation for these data.

Comments below

Although this pie chart isn't entirely wrong, it has two rather silly features which you may have spotted. Firstly, it doesn't make much sense to place a number scale in a circular arrangement, as has been done here. The numbers 0 to 6 fall in a natural sequence and it would be more sensible to find a representation which preserves and even emphasizes this feature – a straight line would be best. Secondly, you may remember

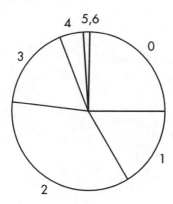

Figure 4.4   Inappropriate use of a pie chart, showing the number of milk bottles on 100 doorsteps

from the bar chart in Figure 4.3 that the slice corresponding to the category '5 bottles' had a frequency of zero. Unfortunately, in a pie chart, any category with a frequency of zero simply disappears. The bar chart has neither of these drawbacks.

However, pie charts do have a special quality which simple bar charts do not.[1] This is that they give a good impression of how large each 'slice' is in relation to the complete pie. In fact, a pie chart should not be used unless the complete pie meaningfully represents the sum of the separate slices. In the case of the contraceptive data (Figure 3.7) the complete pie corresponds to all the different methods of contraception taken together, so it does meet this condition. However, Figure 4.5 shows a pie chart where the condition is not met, and as a result the complete pie chart is meaningless. It has been drawn from the data in Table 4.6.

	DIAMETER (1000 km)
Jupiter	143
Saturn	121
Uranus	51
Neptune	50

Table 4.6   Equatorial diameter of the four largest planets

---

[1] Component bar charts do, however, meet this condition.

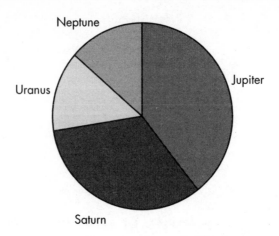

Figure 4.5 A meaningless pie chart drawn from the data in Table 4.6

Stemplots and histograms

We now turn our attention to stemplots and histograms.

EXERCISE 4.8. Analysing stemplots

Have a look at Figures 3.11 and 3.14 and for each of these representations, try to identify

a) *the type of data being represented*
b) *the type of statistical judgement which they might enable you to make.*

Comments below

(a) The data represented in Figure 3.11 were taken from Table 3.3. Table 3.3 is itself a summary of the original raw data, so we have to use our imaginations a little to decide what the original data might have looked like. It is likely that they were a batch of ages, probably measured in whole years. Since age is a continuous variable, it would be reasonable to describe these data as continuous.

However, as was indicated earlier, the data were probably collected in discrete numbers of years, so there is room for ambiguity in this answer. The stemplot data in Figure 3.14 are also taken from the same continuous variable (age), but again it is clear that they have been stated in discrete numbers of years.

(b) Whereas the histogram simply describes the data, the back-to-back stemplot in Figure 3.14 allows you to compare two batches of data side by side.

In general, histograms are only suitable for continuous data, while stemplots can be used for any type of numerical data – discrete or continuous. Individual histograms and stemplots are useful for describing the data, and if you need to make comparisons between two batches of data, then a pair of histograms or a back-to-back stemplot would be needed.

We now turn to the final two types of graph shown in Figures 3.15 and 3.16. These are the scattergraph and the line graph. If you look at these two graphs now, you can see that, in both cases, they have been drawn with two separate variables marked on the axes. It has already been mentioned that these types of graph are useful for representing paired data and allow us to explore inter-relationships.

This completes the description of the common graphs in terms of the sort of data which they are used to represent, and the sort of statistical judgements which they enable you to make. The following final exercise in the chapter will allow you to review and summarize these ideas for yourself.

EXERCISE 4.9. Summary time

Using Table 4.7, summarize the main points of this final section of the chapter. This involves writing in cells of the table the types of graph listed below, bearing in mind the type of data and the type of judgements for which they are normally used.

List of graphs: bar chart, pair of bar charts, compound bar chart, component bar chart, pie chart, pair of pie charts, pictogram, histogram, pair of histograms, stemplot, back-to-back stemplot, scatter-graph, time graph.

TYPE OF DATA	DESCRIBING	COMPARING	INTER-RELATING
Discrete single data			
Continuous single data			
Paired data			

Table 4.7   Type of judgement and type of data together

Comments on page 64

Summary

This chapter has built on the technical description of *how* to draw the various common graphs given in the previous chapter, and has turned to look at *when* they might usefully be drawn. As described here, the choice of graph depends on two crucial factors – the nature of the data being represented (whether they are discrete, continuous, or paired) and also on the sort of statistical judgement which you wish to make about the data (whether you are describing, comparing or inter-relating). The final table which you completed in Exercise 4.9 (and which is completed for you on page 64), should provide you with a useful summary of all these ideas in a single diagram.

Comments on exercises

Exercise 4.1

	WORD CATEGORIES	DISCRETE NUMBERS	CONTINUOUS NUMBERS
Discrete data	✓	✓	
Continuous data			✓
Variables		✓	✓
Attributes	✓		

Exercise 4.2

Data item	Type of data (D or C)
M27	D (word category)
1929*	C
1.2–1.5 m	C
4.5–6.8 kg	C
three buds	D (discrete number)
one quarter	C

Exercises 4.3 to 4.8   Comments in the text.

---

*This is the only item of data in this table which is difficult to classify unambiguously. On the one hand what we have here is a measure of time, in years, and time is a classic text-book example of a continuous measure. However, dates are normally measured in whole numbers of years (we would never talk of the year $1929\frac{1}{2}$, for example) so there would also be a case for calling this a discrete measure.

Exercise 4.9

TYPE OF JUDGEMENT			
TYPE OF DATA	DESCRIBING	COMPARING	INTER-RELATING
Discrete single data	bar chart, pie chart, pictogram, stemplot	pair of bar charts, pair of pie charts, compound bar chart, component bar chart, back-to-back stemplot	
Continuous single data	histogram, stemplot	pair of histograms, back-to-back stemplot	
Paired data			scattergraph, time graph

Note that the stemplot has been placed in both the discrete and the continuous categories in the table above. However, it would only make sense to use a stemplot with a continuous measure if the numbers involved were suitably rounded.

5 SUMMARIZING DATA

We live in a world of rapidly growing collections of data. Information is being amassed and communicated on an ever wider range of human activities and interests. Even if supported by a computer database, a calculator or a statistical package, it is hard for the 'average person' to gain a clear sense of what this information might be telling them. A crucial human skill is to be selective about the data that we choose to analyse and, where possible, to summarize the information as briefly and usefully as possible. In practice, the two most useful questions which will help you to summarize a mass of figures are:

What is a typical, or average value?

and

How widely spread are the figures?

This chapter looks at a few of the most useful summary measures under these two headings, *average* and *spread*.

Introducing averages – finding a middle value

The next section looks in detail at three particular averages and how they are calculated. First, however, it is worth focusing on what an average is and why we might bother to calculate it. As was suggested earlier, it is often difficult when looking at a mass of figures to 'see the wood for the trees'. A sensible strategy is to find some typical or middle value which may be taken as being representative of the rest. However, there are several possible ways of finding a representative value, as the following simple exercise illustrates.

EXERCISE 5.1. Typical of kids

Table 5.1 gives the names of nine female friends and the number of children they have.

PERSON	CHILDREN
Roberta	0
Alice	3
Rajvinder	0
Jenny	1
Fiona	2
Sumita	2
Hilary	3
Marcie	0
Sue	4

Table 5.1   Numbers of children of nine women

How would you summarize this information with a single representative number? See if you can think of several plausible ways of doing this.

Comments below

Here are three possible ways of summarising the numbers.

■ One possible approach which you may have chosen is to ask what is the most common number of children. This can be quickly checked out in your head, and your thinking might have involved reorganizing the data as follows.

Number of children	0	1	2	3	4
Frequency	3	1	2	2	1

This reveals that the most frequently occurring number of children is zero (corresponding to a frequency of 3), so one possible single summary is:

Average = 0.

This sort of average, the most frequently occurring number, is called the *mode*.

■ A difficulty with using the mode in this example is that it has come up with a value which could not be described as 'typical' in the sense of lying somewhere in the middle of the range of possible values from 0 to 4. One way of achieving a middle value is by adding the nine numbers together and dividing by nine, thus:

$$\frac{0 + 3 + 0 + 1 + 2 + 2 + 3 + 0 + 4}{9} = \frac{15}{9} = \frac{1.667}{\text{children}}$$

This type of average is often called the 'average', but clearly this is not a sufficiently precise term since there are several types of average around. The 'correct' statistical term for this type of average is the *mean*.

■ The mean has at least given us a representative value which lies within the range of possible values from 0 to 4. However, since children are usually measured in whole units, it has produced a number which can hardly be described as typical! This example seems to demand that we try to find a representative value which both lies within the range 0 to 4 and also gives a whole number. Fortunately there is a measure which does just that. What we need to do is to sort the nine numbers into order from smallest to largest and then simply select the middle number in the sequence. For nine numbers, the middle one will be the fifth (giving four numbers on either side). Thus:

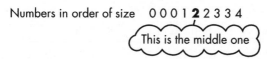

Numbers in order of size   0 0 0 1 **2** 2 3 3 4

This is the middle one

By this method, we get a representative value of 2 children. This type of average is called the *median*. When there is an odd number of items, as there is here, it is clear that there will be a single median value. However, suppose that there had been an even number of values. For example, let us exclude Sue's four children from the group and try to find the median number of children from these eight numbers.

Numbers in order of size 00 0 0 **1 2** 2 3 3

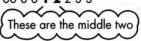

These are the middle two

There are now two middle values – the fourth and fifth numbers. In order to find the median, we must now find the mean of these two middle values.

Thus, in this case, median $= \dfrac{1 + 2}{2}$ or 1.5.

In this particular example, the median would seem to be the most suitable sort of average. However, it has to be admitted that this has been a fairly artificial example. Since the choice of average really depends on context and on the purpose for calculating it in the first place, this wouldn't be a good illustration of how you might go about selecting a suitable average for a given problem. Some specific suggestions are given in the next section.

Calculating and choosing averages

We now look in a bit more detail at how these three averages – the mode, the mean and the median – are calculated and also at how we can decide which one to choose under which circumstances.

Mode

The mode of a batch of data is usually defined as the most frequently occurring item. The procedure for finding the mode is as follows:

- *(i)* identify all the distinct values or categories in the batch of data;

- *(ii)* make a tally of the number of occurrences of each value or category;

- *(iii)* the mode is the value or category with the greatest frequency.

Here are three simple examples to give you practice at calculating the mode for different types of data.

EXERCISE 5.2. Finding the mode

Three additional sets of data are supplied in Table 5.2, all associated with the nine friends listed in Table 5.1. Respectively, these data sets can be described as (a) category data (eye colour), (b) discrete numerical data (lucky number) and (c) continuous numerical data (height).

Calculate the mode for each of these data sets.

NAME	(a) EYE COLOUR	(b) LUCKY NUMBER	(c) HEIGHT (m)
Roberta	blue	3	1.56
Alice	green	7	1.62
Rajvinder	brown	7	1.71
Jenny	brown	9	1.58
Fiona	blue	3	1.67
Sumita	brown	5	1.66
Hilary	green	7	1.60
Marcie	brown	7	1.59
Sue	blue	7	1.61

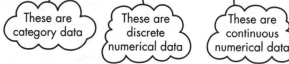

Table 5.2   Three different data types

Comments on page 88

There is a problem when calculating the mode from continuous numerical data, of the type in column (c) of Table 5.2. The difficulty, and one that is in evidence here, is that each number is different. As a result, they each have a tally of 1, which makes the mode – the most frequently occurring value – rather meaningless. The problem can be solved by grouping the nine values into intervals of 5 cm, as shown in Table 5.3.

When the data have been grouped, it is possible to point to the interval with the greatest frequency and this is called the 'modal interval' – in this case the modal interval is $1.55 < H \leqslant 1.60$, with a frequency of four.

	HEIGHT (H) M	TALLY	
This is the modal interval . . .	1.55 < H ≤ 1.60 1.60 < H ≤ 1.65 1.65 < H ≤ 1.70 1.70 < H ≤ 1.75	IIII II II I	*. . . because this is the largest frequency*

Table 5.3   Tally of heights in 5 cm intervals

Overall, the mode is particularly useful for categorical data or discrete numerical data but can only be used meaningfully with continuous data if they are grouped into intervals, as has been done here.

A simple mean ($\bar{X}$)

The mean, or the arithmetic mean, as it is sometimes called, is the best known average which can be defined as 'the sum of the values divided by the number of values'. The mathematical symbols used to describe the mean were explained in Chapter 2. Here they are again.

the symbol for the mean, pronounced 'X bar' $$\bar{X} = \frac{\Sigma X}{n}$$ *the Greek letter capital sigma, meaning 'the sum of'*

EXERCISE 5.3. Finding the mean

Try to find the mean of the three sets of data in Table 5.2.

Comments below

■ If you didn't fail to find the mean of data set (a), you should have! It is simply nonsensical to find the mean of categorical data.

■ It is possible to calculate the mean of the nine numbers in data set (b). This gives:

$$\bar{X} = \frac{3 + 7 + 7 + 9 + 3 + 5 + 7 + 7 + 7}{9} = \frac{55}{9} = 6.111 \ldots$$

Wow – some lucky number! Clearly, although a mean can be found here, the number you get, 6.111 ..., is not remotely useful or interesting, so I'm sorry to have wasted your time! Perhaps the lesson to learn from this exercise is that the mean is a somewhat overworked average, and you need to question whether it meets the main criterion of any well-chosen average; namely, does it provide a useful and representative summary of the data set? If it doesn't, then don't use it.

■ Continuous data of the type given in data set (c) are usually ideal for calculation of the mean. The calculation in this case is:

$$\bar{X} \frac{1.56 + 1.62 + 1.71 + 1.58 + 1.67 + 1.66 + 1.60 + 1.59 + 1.61}{9}$$

$$= \frac{14.60}{9} = 1.622 \text{ m.}$$

Note that it is usual to give the answer to the calculation of a mean to one place of decimals more than the original data, as has been done here (three decimal places compared with two in the original heights).

A weighted mean ($\bar{X}$)

We shall now look at what is involved when calculating means of data sets with much larger sample sizes and where the data are organized either in frequency tables or percentage tables, depending on which is the more appropriate. For example, Table 5.4 shows the distribution of households by size for 1961 and 'today' for a large city in the UK. It simply wouldn't make sense to write each number out separately for

HOUSEHOLD SIZE	1961	'TODAY'
1	12	25
2	30	34
3	23	17
4	19	16
5	9	6
6+	7	2

Table 5.4   Households by size (percentages) in a large city in the UK
Source: Adapted from national data

every person surveyed. Instead, as shown in Table 5.4, the data have been grouped and, for convenience, are stated in percentages. The figures for 1961 and 'today' are given side by side.

Seeing the 1961 and 'today' distributions of household sizes side-by-side makes for an interesting comparison. Two significant differences between these two distributions occur at the extremes – i.e. at household size 1 and 6+. The proportion of single-person households 'today' has roughly doubled compared with 1961 (25% compared with 12%), while the proportion of large households containing 6+ people dropped from 7% to 2%. Overall, then, there has been a significant shift to smaller household sizes and this should be reflected in a reduction in the average household size, as the calculation of the mean below will show.

With this example, the basic definition of the mean as 'the sum of the values divided by the number of values' will not give the correct answer. The particular complication here is that the different household sizes, 1, 2, 3 etc, do not occur with the same relative frequency. Taking the 1961 figures, for example, 12% of the households were of size 1, whereas 30% were of size 2. So clearly these different-sized households are not equally weighted and any averaging procedure would need to take account of the unequal weights. A second complication occurs with the final household size, 6+. This could include households with 7, 8, 9 etc people, right up to 20 or 50 or maybe even 100! Of course, there will be many more households of size 6 or 7 than there will be, say, of size 20. What we have here, then, is an interval where the upper limit has not been specified. What we need to do is take a sensible number of people, say 9, which would be representative of this interval. So, how do we come to the value 9? Well, without further information, the precise value chosen is just a matter of common sense and intelligent guesswork.

We shall now calculate the mean household size for 1961. This time we shall use the procedure for calculating a 'weighted mean', which involves the following stages.

(i) multiply each household size (X) by its corresponding 'weight' or frequency (f). In this example, the 'f' refers to the percentage of households of household size 1, 2, 3, etc.

(ii) add these products together and divide by the sum of the weights.

HOUSEHOLD SIZE (X)	WEIGHT (f)	PRODUCT (fX)
1	12	12 × 1 = 12
2	30	30 × 2 = 60
3	23	23 × 3 = 69
4	19	19 × 4 = 76
5	9	9 × 5 = 45
6+ (assume = 9)	7	7 × 9 = 63
TOTAL	100	325

this is the sum of the weights

this is the sum of the fX products

Table 5.5   Calculating a weighted mean for the 1961 data on household size

Referring to Table 5.5, weighted mean = $\dfrac{325}{100}$ = 3.25 persons per household.

EXERCISE 5.4. Calculating a weighted mean yourself

Now you have seen the calculation of the weighted mean for the 1961 figures, use the same approach to calculate the weighted mean for the 'today' figures.

Comparing the two results, what do they suggest about how average household size has altered over the period in question?

Comments below

The weighted mean for 'today' is 2.52 people per household, which, as we predicted, is substantially less than the 1961 figure of 3.25.

Finally, here is a formal mathematical description of how a weighted mean is calculated.

the mean, 'X bar' $\bar{X} = \dfrac{\sum fX}{\sum f}$ this is 'sigma fX', the sum of the separate products of each X value with its corresponding frequency. i.e. $f_1X_1 + f_2X_2 + f_3X_3 +$ etc

this is 'sigma f', the sum of the frequencies, $f_1 + f_2 + f_3 +$ etc

Median (Md)

As was shown earlier, the median of a set of, say, nine values is found by ranking the values in order of size and choosing the middle one – in the case of nine values it is the value of the fifth one. Note that a common mistake is to say that the value of the median of nine items is 5. It needs to be stressed that the median is *the value* of item number 5 when all nine items are ranked in order and is not its rank number. With an odd number of items, like 11 or 37 or 135, there will always be a unique item in the middle (respectively the 6th, 19th and 68th). A simple way of working this out is to add one to the batch size and divide by 2. For example, for a batch of 37 items, $(37 + 1)/2 = 19$, so the median is the value of the 19th item. However, where the batch size of the data is even, there is not a unique item in the middle. In such circumstances, the usual approach is to choose the two middle values and find their mean.

Example　A count of the contents of eight boxes of matches produced the following results:

> Number of matches　52　49　47　55　54　51　50　50

Find the median number of matches in a box.

Solution　Ranking the numbers in order of size produces the following:

47　49　50　50　51　52　54　55

the two middle values

Since there is an even number of values, the median is the mean of the two middle values.

i.e. median $= \dfrac{50 + 51}{2} + \; = 50.5$

Now clearly boxes of matches do not tend to be bought with fractions of a match in the box, so in this example, and indeed in general, the median needs to be interpreted with care.

Let us now look at how we calculate the median from a frequency or percentage table. The stages are as follows.

　　(i)　Redraw the table as a cumulative percentage table.

　　(ii)　Identify the value corresponding to the 50% position (remember that the median can be found 50% of the way

along the values when they have been ranked in order of size).

Example Find the median household size for the 1961 data in Table 5.4.

Solution Table 5.6 shows the data extended to include a cumulative percentage table.

HOUSEHOLD SIZE (S)	PERCENTAGE (P)	HOUSEHOLD SIZE (≤S)	CUMULATED PERCENTAGE (P)
1	12	1	12
2	30	2 or less	42
3	23	**3 or less**	**65**
4	19	4 or less	84
5	9	5 or less	93
6+	7	all	100
TOTAL	100		

e.g. 65 = 12 + 30 + 23

Table 5.6 Cumulative percentage table showing the 1961 data on household size

The highlighted line in the cumulative percentage part of Table 5.6, corresponding to a household size of '3 or less', covers the 23 percentage points ranging from 43% to 65% inclusive. Clearly the 50% value lies in this range, so '3' must be the household size that contains the median.

EXERCISE 5.5. Calculating a median yourself

Now that you have seen the calculation of the median for the 1961 figures, use the same approach to calculate the median for the 'today' figures (also given in Table 5.4).

Comparing the two results, what do they suggest about how average household size has altered over the period in question?

Comments on page 88

Finally, to end this section on averages, Table 5.7 provides a summary of the main characteristics of the three averages and the sort of data they can usefully summarize.

TYPE OF AVERAGE	TYPE OF DATA		
	Category	**Discrete numerical**	**Continuous numerical**
Mode	Ideal – only the mode will do for most category data	Fine, provided some of the values occur more than once	Only if the data are grouped into intervals
Mean	No good	Fine, but be careful to check that the answer is sensible	Good
Median	Only if the data have a natural order	Fine, but be careful to check that the answer is sensible	Good

Table 5.7   An average for all data types

Spread

As you have seen from the previous section, knowing the approximate centre of a batch of data is a useful and obvious summary of the entire batch. The second key question which we posed at the beginning of the chapter was 'How widely spread are the values?' This section deals with the following five measures of spread:

- range
- inter-quartile range
- mean deviation
- variance
- standard deviation

Since people are increasingly performing statistical calculations like these with the aid of a calculator or suitable computer package (perhaps a spreadsheet or statistical package), treatment of these techniques here will focus less on the mind-numbing arithmetic that can be involved and more on the principles on which they are based. To this end, a very elementary data set has been used throughout, simply to lay out the bare bones of each calculation. The data are the heights of the nine women first given in Table 5.2 and they are repeated in Table 5.8.

NAME	HEIGHT (m)
Roberta	1.56
Alice	1.62
Rajvinder	1.71
Jenny	1.58
Fiona	1.67
Sumita	1.66
Hilary	1.60
Marcie	1.59
Sue	1.61

Table 5.8   The heights of nine women

5-figure summary

Before looking at the range and inter-quartile range, it is necessary to scan these nine numbers in order to pick out five of the key figures. This is easiest to do when the numbers are sorted in order of size, as shown in Table 5.9.

- The lower extreme value (E_L) is the height of the shortest person, Roberta, with a height of 1.56 m.
- The upper extreme value (E_U) is the height of the tallest person, Rajvinder, with a height of 1.71 m.
- The median value (*Md*) lies half way through the values in Table 5.9. This corresponds to Sue's height, so $Md = 1.61$ m.
- The lower quartile (Q_L) lies a quarter of the way through the values in Table 5.9.

NAME	HEIGHT (m)	FIVE KEY FIGURES
Roberta	1.56 ⟶	The lower extreme value (E_l)
Jenny	1.58	
Marcie	1.59 ⟶	The lower quartile (Q_l)
Hilary	1.60	
Sue	1.61 ⟶	The median height (*Md*)
Alice	1.62	
Sumita	1.66 ⟶	The upper quartile (Q_U)
Fiona	1.67	
Rajvinder	1.71 ⟶	The upper extreme value (E_U)

Table 5.9   The heights of nine women, sorted in order of size

To put this another way, the lower quartile lies half way through the lower half (i.e. covering the five shortest people) of the batch of data. Marcie is third of these five women, so $Q_L = 1.59$ m.

■ The upper quartile (Q_U) lies three quarters of the way through the values in Table 5.9. By the same reasoning as before, this will correspond to the median height of the five tallest women. This will be Sumita's height, so $Q_U = 1.66$ m.

These five numbers provide a useful summary of a batch of data and are often written in the following arrangement, known as a '5-figure summary'.

$$
\begin{array}{c|cc}
 & Md & \\
 & Q_L & Q_U \\
n & & \\
 & E_L & E_U \\
\end{array}
$$

Included to the left of the 5-figure summary is the sample size, 'n', also known as the batch size or the 'count'. The 5-figure summary for these data is given in Figure 5.1.

$$
\begin{array}{c|cc}
 & 1.61 & \\
 & 1.59 & 1.66 \\
9 & & \\
 & 1.56 & 1.71 \\
\end{array}
$$

Figure 5.1 Five-figure summary for the heights of nine women (m)

A very effective way of picturing the information in a 5-figure summary is to use a *boxplot* (sometimes called a 'box-and-whisker' plot).

Boxplot

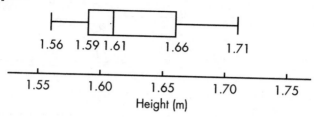

Figure 5.2 Boxplot drawn from the 5-figure summary in Figure

In general, then, the significant parts of the boxplot are as follows.

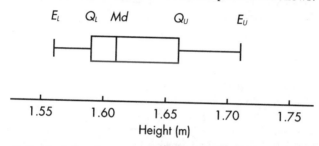

Figure 5.3 Boxplot showing where the 5-figure summary values are placed

The central rectangle which marks out the two quartiles is called the 'box', while the two horizontal lines on either side are the 'whiskers'. Just by observing the size and balance of the box and the whiskered components we can gain a very quick and useful overall impression of how the batch of data is distributed. Also, drawing two boxplots one above the other can provide a powerful means of comparing two distributions. For example, have a look at Table 5.10 which compares wing lengths, in

	FEMALES	MALES
Highest decile	82	85
Upper quartile	80	83
Median	77	81
Lower quartile	76	80
Lowest decile	74	78

Table 5.10 Wing lengths, in millimetres, of meadow pipits

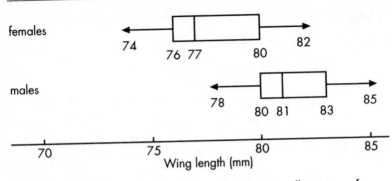

Figure 5.4   Decile boxplots of wing lengths, in millimetres, of
meadow pipits
Source: Adapted from Open University course 'Using Mathematics,
MST121', Chapter D4

millimetres, of two very large samples of male and female meadow
pipits. The corresponding boxplots are in Figure 5.4.

EXERCISE 5.6 Comparing boxplots

**From the boxplots above, what general features can
you pick out about wing lengths of the male and
female meadow pipits?**

Comments below

There are several clear patterns here, of which four are mentioned below.

■ Notice that the 'whiskers' end with arrow heads, rather than
short vertical lines. This is because these end points are
'deciles' (the lowest decile cuts off the bottom 10% and the
highest decile cuts off the top 10% of the values) rather
than the lower and upper extreme values. With very large
batches (these data are based on a large batch size of birds)
it really isn't very relevant to know what the extreme values
are and so we tend to use the deciles instead.

■ The wing lengths of the male birds are noticeably larger
than those of the females. For example, the upper quartile

of the female wing length coincides with the lower quartile of those of the males.

■ Female wing lengths are slightly more widely spread than those of the male. Compare the length of the boxes between female and male and observe that the interval between the quartiles for female wing lengths is slightly wider. The same is true of the intervals between the arrow heads (the inter-decile range).

■ Both boxplots are slightly *skewed* to the right – notice how both of the right hand parts are slightly longer than their equivalent left hand parts. Incidentally, the direction by which skewness is described – skewed to the right or to the left – may seem counter-intuitive. It may help you to remember that the side which has the long tail is the direction to which we say it is skewed.

Range

The range is the simplest possible measure of spread and is the difference between the upper extreme value (E_U) and the lower extreme value (E_L). Returning to the height example, the tallest person is Rajvinder at 1.71 m and the shortest is Roberta at 1.56 m. Thus:

Range = $E_U - E_L$ = 1.71 m − 1.56 m = 15 cm.

One disadvantage of the range as a measure of spread is that it is strongly affected by an extreme or untypical value. For example, it only takes one extremely tall or extremely short person in the sample to have an enormous effect on the value of the range. This problem can be overcome by choosing to measure the range, not between the two extreme values, but between two other values on either side of the middle value. And what could be two more appropriate values to choose than the upper and lower quartiles!

Inter-quartile range (dq)

The inter-quartile range, sometimes known as the inter-quartile deviation, is, as you might expect, the difference between the two quartiles. In this example it is the difference between the heights of Sumita and Marcie.

The inter-quartile range, $dq = Q_U - Q_L$ = 1.66 m − 1.59 m = 7 cm.

The inter-quartile range therefore contains the middle half of the batch values, which is, of course, the central box part of the boxplot.

A difficulty with any measure of spread which is tied to the units of the batch values is that it makes comparisons with the spread of other batches of data very misleading. For example, how might you compare the spread of a group of people's heights, measured in metres, with the spread of their weights, measured in kg? And would the spread of heights suddenly increase by a factor of 100 if the data values were converted from metres to centimetres? Even comparing the spreads of two batches measured in the same units can be misleading, as the next exercise shows.

EXERCISE 5.7 Comparing spreads

Using the data from Table 5.10, calculate the inter-quartile ranges for male and female wing lengths of the meadow pipits. What does this reveal about how widely spread their wing lengths are?

Comments below

---

dq for female wing lengths $80 - 76 = 4$ mm

dq for male wing lengths $83 - 80 = 3$ mm

We can see that the dq value for females, 4 mm, is slightly more than the dq value for males of 3 mm. So, on the face of it, female wing lengths seem to be more widely spread than those of males. However, this is a slightly misleading conclusion because we are not comparing like with like. Since male wing lengths are slightly more than females', we would expect any calculation based on male wing lengths to produce a larger result than a corresponding calculation based on female wing lengths. What we need to do is to *standardize* the inter-quartile range so that its value is independent of the actual batch values. This is done by expressing each dq value as a proportion of its corresponding median value, thus:

Standardized inter-quartile range $= \dfrac{dq}{Md}$

Standardized inter-quartile range for female wing lengths $= \dfrac{4}{77} = 0.052$

Standardized inter-quartile range for female wing lengths $= \dfrac{3}{81} = 0.037$

This standardized result suggests that, when expressed as a proportion of the median, the spread of female wing lengths is, in fact, even greater than that of males. Another example of standardizing the spread is provided in the next chapter.

Mean deviation

The mean deviation is a fairly commonsense measure of spread. It can be described as the 'average deviation of each batch value from the mean'. You will see how this measure is worked out in practice, but it will be introduced first with a deliberate mistake. As you work through the calculation, see if you can spot the error!

Here are the main stages involved in calculating the mean deviation. (Watch out for some silliness on the way.)

(i)   Calculate the overall mean.

(ii)   Subtract the mean from each value in the batch. This produces a set of deviations, d (where $d = X - \bar{X}$ for each value of X).

(iii)   Find the mean of these deviations. The result is the 'mean deviation'.

Table 5.11 shows how these instructions apply to the height data.

As you can see, the calculation as it has been done here has produced an answer of zero for the mean deviation. Clearly, since there *is* a spread around the mean, this cannot be correct. So what has gone wrong?

The explanation lies with a basic property of the mean, namely that the sum of the positive deviations is exactly matched by the sum of the negative deviations. Therefore, the deviations *must* sum to zero – if they

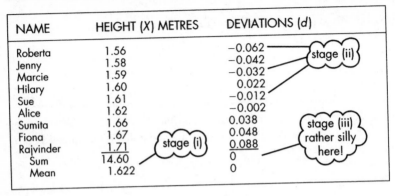

Table 5.11   Incorrect calculation of the mean deviation

didn't, you would know that you had made a mistake! The problem is solved by treating all the deviations as positive. The mean deviation can now be defined more properly as the 'mean of the modulus of the deviations around $\bar{X}$' (where the term 'modulus' means 'the positive value of'). This is written in symbols as follows.

Mean deviation, MD, $= \dfrac{\sum |d|}{n}$ — said as 'sigma modulus d over n'. The modulus is written with two vertical lines, as shown here

Here, now, are the correct instructions for calculating the mean deviation.

(i)   Calculate the overall mean.

(ii)   Subtract the mean from each value in the batch. This produces a set of deviations, d (where $d = X - \bar{X}$ for each value of X). Take the modulus of these deviations, $|d|$.

(iii)   Find the mean of these positive deviations – giving the 'mean deviation'.

Table 5.12 now shows how to calculate the mean deviation correctly!

Variance (σ^2) and Standard deviation (σ)

The reason that the mean deviation (above) was introduced via a deliberate error, apart from being a desperate attempt to keep the reader awake,

NAME	HEIGHT (X) METRES	MODULUS OF THE DEVIATIONS ($|d|$)
Roberta	1.56	0.062
Jenny	1.58	0.042
Marcie	1.59	0.032
Hilary	1.60	0.022
Sue	1.61	0.012
Alice	1.62	0.002
Sumita	1.66	0.038
Fiona	1.67	0.048
Rajvinder	1.71	0.088
Sum	14.60	0.347
Mean	1.622	0.385

stage (ii) (pointing to 0.042, 0.032, 0.022, 0.012)
stage (iii) (pointing to 0.088)
stage (i) (pointing to 14.60)

Table 5.12   Mean deviation, MD = 0.0385 metres, or 3.85 cm.

was in order to raise the question of how we should deal with the fact that the sum of the deviations around the mean is zero. The problem was solved for the *MD* simply by making all the negative numbers positive. An alternative way of turning negative numbers into positive ones is to square them. This is the approach used when calculating the variance and the standard deviation. These two measures of spread are closely linked – the variance is the square of the standard deviation.

The symbol for variance is σ^2, while standard deviation is written as σ. The letter σ is the lower case form of the Greek letter 'sigma', not to be confused with the upper case version, Σ, which means 'the sum of'.

The method of calculating the variance is described below. As you read the description, glance back to how the mean deviation is calculated. Notice what is the same and what is different in the two explanations.

   (i)   Calculate the overall mean.

   (ii) Subtract the mean from each value in the batch. This produces a set of deviations, d (where $d = X - \bar{X}$ for each value of X). Take the square of these deviations, d^2.

   (iii) Find the mean of these squared deviations, $\dfrac{\Sigma d^2}{n}$. The result is the 'variance'.

(iv) If you wish to find the standard deviation, take the square root of the variance.

Now, here are the formulas for these two measures of spread.

Variance, $\sigma^2 = \dfrac{\Sigma d^2}{n}$

Standard deviation, $\sigma = \sqrt{\dfrac{\Sigma d^2}{n}}$

hw

Finally, Table 5.13 shows how to calculate the variance and standard deviation, using the height data.

NAME	HEIGHT (X) METRES	DEVIATIONS (d)	SQUARED DEVIATIONS (d^2)
Roberta	1.56	−0.062	0.003844
Jenny	1.58	−0.042	0.001764
Marcie	1.59	−0.032	0.001024
Hilary	1.60	0.022	0.000484
Sue	1.61	−0.012	0.000144
Alice	1.62	−0.002	0.000004
Sumita	1.66	0.038	0.001444
Fiona	1.67	0.048	0.002304
Rajvinder	1.71	0.088	0.007744
Sum	14.60	0	0.018756
Mean	1.622	0	0.002084
			0.045651

stage (ii)

stage (iii), variance

stage (i)

stage (iv), standard deviation

Table 5.13   Calculation of the variance and standard deviation

So, variance = 0.002 084 m^2 and standard deviation = 0.045 651 m. If this were a practical example, these results would, of course, be rounded to an appropriate number of decimal places, depending on the circumstances of the question.

The procedure for calculating the variance and standard deviation is slightly more complicated when applied to data that have been organized into a frequency or percentage table. However, the method is exactly equivalent to the approach used when calculating a weighted arithmetic mean; namely, you have to remember to multiply each value (in this case, each squared deviation) by its corresponding weight, f, before finding the sum of the squared deviations. The formulas for finding a weighted variance and standard deviation are given below.

Variance, $\sigma^2 = \dfrac{\Sigma f d^2}{\Sigma f}$

where $d = X - \bar{X}$ for each value of X

Standard deviation, $\sigma = \sqrt{\dfrac{\Sigma f d^2}{\Sigma f}}$

remember that $\Sigma f = n$

There are other formulas for calculating the variance and standard deviation which will produce the same results but which make for a slightly easier calculation. For reasons of space these are not explained here, but increasingly these short-cut methods are becoming irrelevant as more and more users are performing such calculations on machines.

Finally, it is worth noting that of all the measures of spread that are available, the standard deviation is the one most widely used.

Summary

This chapter dealt with a variety of statistical techniques for summarizing data. Firstly, under the general theme of finding a typical or average value, we looked at the mean, the mode and the median.

The second key question central to summarizing data was, 'how widely spread are the figures?' and this led to the following measures of spread – 5-figure summary, boxplot, range, inter-quartile range, mean deviation, variance and standard deviation.

Comments on exercises

Exercise 5.2

    (a) The mode is 'brown', with four occurrences.

    (b) The mode is '7', with five occurrences.

    (c) As is explained in the main text, calculating the mode with these data is meaningless.

Exercise 5.5

The median for 'today' is 2 people per household, which is less than the 1961 figure.

6 | LIES AND STATISTICS

It's not the size of your statistic, it's what you do with it that counts.

A variety of distortions, both deliberate and innocent, are commonly employed when numbers are used to bolster an argument. There is much scope in statistics for a little 'creativity', particularly when it comes to beguiling with a graph, being less than accurate with averages or even being perfidious with a percentage. The three sections of this chapter deal with these three potentially misleading areas of statistics – graphs, percentages and averages.

Misleading graphs

The first and most important thing to remember about drawing sensible graphs is that you need to be clear about the *purpose* for which the graph is to be drawn. The implication of this is that there is no single 'correct' graph for a particular data set – it all depends on what point you want the graph to make for your data. Having said that, there are a number of situations when a particular graph is clearly wrong for the type of data it has been used to represent and some examples are given here.

This section on graphs is based on four data sets, A, B, C and D and a variety of possible graphical representations is offered for each. Some of the graphs suggested are quite sensible, while others are plain daft! Comments are given at the end of each collection of graphs but before reading them, spend some time yourself looking at the suggested graphs and try to decide which are sensible and which are silly.

Data set A

NAMES	TIMES
Anthony	26
Emma	18
Jaspal	19.6
Lisa	21
Meena	22
Navtej	27
Nicola	23
Sandeep	17
Tanya	23
Thomas	19

Table 6.1   Data set A: times (seconds) for the egg and spoon race at Cookstown village fete

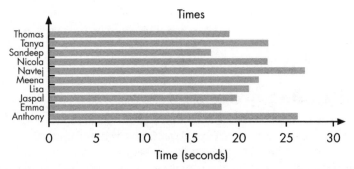

Graph A1   Horizontal bar chart showing the data in Table 6.1

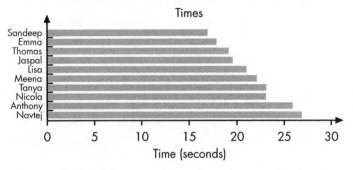

Graph A2   Ordered horizontal bar chart showing the data in Table 6.1

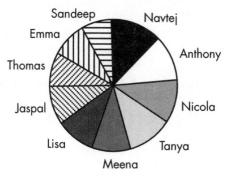

Graph A3 Pie chart showing the data in Table 6.1

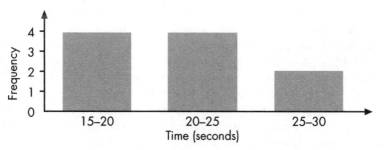

Graph A4 Block graph showing the data in Table 6.1

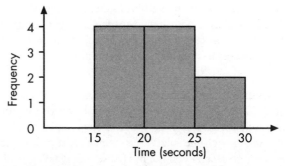

Graph A5 Histogram showing the data in Table 6.1

Comments below

First a word about Data set A. The first column shows a set of names, so this is a set of *discrete* data. The second column gives running times and time is a *continuous* measure. What sort of graph you choose here will depend on whether you want the data to be organized principally on the basis of the names or the times. In general, discrete data are depicted using a bar chart or pie chart, while continuous data are best represented with a histogram.

Graphs A1 and A2 show this data set represented in bar charts where the bars have been drawn horizontally rather than in the usual vertical orientation. Drawing bar charts this way round is a useful device when there is a large number of bars and there would be too much writing to put the labels onto the horizontal axis. Graph A2 is more helpful than A1 here, since it has placed the bars in rank order, showing clearly that Sandeep is the quickest and Navtej the slowest with an egg and spoon. However, both these graphs have the drawback that they don't show the distribution of data.

Graph A3 is a complete disaster! Although pie charts are suitable for discrete data (as this data set is), and each slice of this pie does correspond to a separate 'discrete' category, the second important condition for using a pie chart has not been met here. This is that the complete pie needs to represent something which is complete and meaningful. This complete pie, however, corresponds to the sum of all the running times and this is not a particularly useful or interesting collection of things.

A different approach is taken in graphs A4 and A5, where the focus of interest is overall patterns in the running times and the people's names have been ignored. These graphs have been produced as a result of grouping the times into 5 second intervals, counting the frequencies in each group, and drawing them as a block graph. A weakness of graph A4 is that it has been drawn with gaps between adjacent columns. As was explained in Chapter 4, this practice should only be used when depicting discrete data, whereas the variable, time, is continuous.

Verdict

The most favoured graphs here are A2 and A5 but they each do a quite different job. A2 preserves the raw data and allows you to compare the relative times of each person in the race. A5 gives a picture of the data as a whole and shows the time intervals in which most or least people fall. Whichever of these graphs you might choose will depend on the purpose for which you wished to draw the graph in the first place.

Data set B

WIND TYPE	DAYS
Strong wind	10
Calm	5
Gale	7
Light breeze	9
Total	31

Table 6.2   Data set B: wind in January

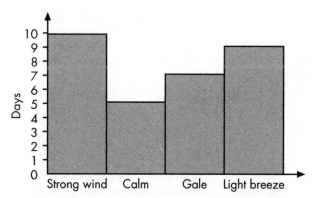

Graph B1   Block graph showing the data in Table 6.2

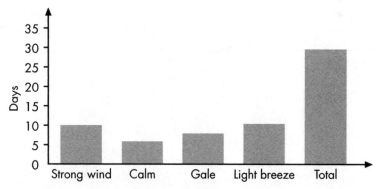

Graph B2   Bar chart showing the data in Table 6.2

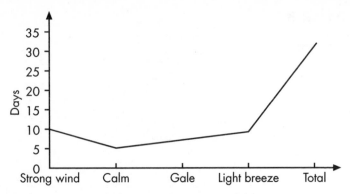

Graph B3 Line graph showing the data in Table 6.2

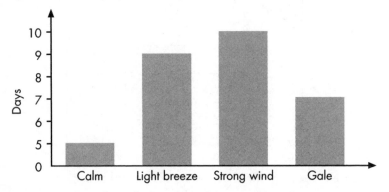

Graph B4 Bar chart showing the data in Table 6.2

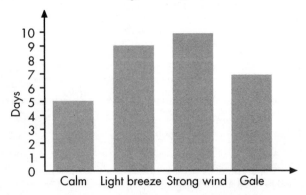

Graph B5 Bar chart showing the data in Table 6.2

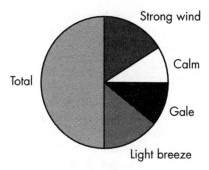

Graph B6   Pie chart showing the data in Table 6.2

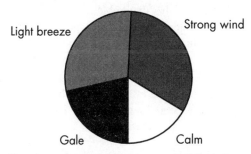

Graph B7   Pie chart showing the data in Table 6.2

Comments below

Data set B is *discrete* and shows the 31 days of a particular January cate-gorized into four headings of windiness. Two features of this table of data are worth noting as they caused problems in some of the subsequent graphs. The first is the inclusion of the 'Total' category and the second is the fact that no attempt has been made to present the headings in any logical order.

Graph B1 contains two problems. Firstly, the sequence of wind cat-egories is the same unhelpful one given in the table above it. A more logical ordering might be calm, light breeze, strong wind and gale. Sec-ondly, the adjacent bars have been drawn so that they touch and this is not appropriate for discrete data.

Graph B2 has at least solved the problem of touching bars by drawing them with gaps but the sequence of the categories is still unsatisfactory.

Also the final 'Total' column which has been added is confusing as it implies that there are five categories when in fact the fifth is simply the sum of the first four columns.

All of these drawbacks apply to graph B3 but there is an additional twist. Joining the tops of the bars together here is meaningless for two reasons. Firstly such a technique is only sensible for continuous data and should therefore be reserved for situations where each point on the joining line or curve actually means something. For example, suppose you plotted a child's height on two successive Januaries. The line joining these points gives a reasonable estimate as to that child's height over the intervening months. This approach is not suitable for discrete data where it is hard to imagine what intervenes between the separate categories. This raises the second issue here, which is that the problem is made worse by the illogical order in which the categories have been presented. As a result, any pattern in the shape of the graph is totally spurious.

Most of these problems have been solved in graphs B4 and B5. However, there is a problem with B4. Note how the vertical axis effectively begins at 4. This has a rather distorting effect on the columns. For example, the 'strong wind' column has been drawn to be six times as tall as the 'calm' column when in fact it only has twice the frequency (10 compared with 5). While it is not 'illegal' to draw the vertical axis so that it starts with a non-zero value, the usual practice is to show a 'break' in the axis as a way of alerting the reader to the potentially misleading interpretation. This is shown below.

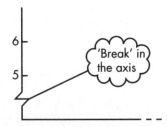

And finally to the two pie charts. The good news is that this data set is highly suitable for a pie chart because it satisfies the two basic conditions, namely that the data are discrete (yes, they are) and that, taken as a whole, the complete pie means something. This second condition certainly holds for graph B7 since the whole pie represents the 31 days of

January. However, graph B6 is a bit daft since the 31 days have been included twice – once in the four small slices and then again in the total.

Verdict

Only two of these graphs appear to be satisfactory, the bar chart B5 and the pie chart B7.

Data set C

Margarine	6 oz
Self-raising flour	1 lb
Caster sugar	6 oz
Sultanas	2 oz
Vanilla essence	$\frac{1}{4}$ tsp
Eggs	3

Table 6.3   Data Set C: Queen Cakes ingredients

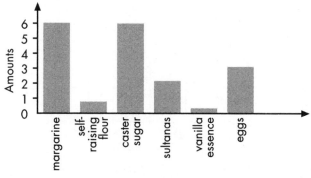

Graph C1   Bar chart showing the data in Table 6.3

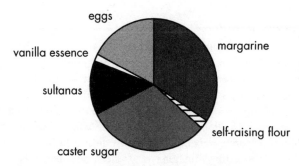

Graph C2   Pie chart showing the data in Table 6.3

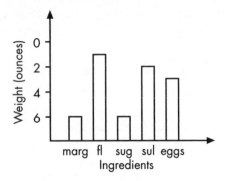

Graph C3 Bar chart showing the data in Table 6.3

Graph C4 Bar chart showing the data in Table 6.3

Comments below

One characteristic of recipe ingredients is that they tend to be measured in different units. For example, weights may be in ounces, pounds, grams and so on. Some ingredients aren't weighed at all – eggs for example – while sometimes non-standard measures are used – teaspoons, cups and so on. As a result, the *numbers* used to measure the different ingredients *are not comparable with each other*. Since the whole point of graphs is to be able to compare numbers visually, this gets Data Set C off to a bad start. Indeed, it is hard to imagine why one might ever wish to graph these data!

Because of this problem of differing units, it is a meaningless exercise to compare the heights of the columns in the bar chart (graph C1) and assess the relative sizes of the slices of the pie chart (graph C2). For example, according to the graphs, you can see that this cake appears to be seriously under-supplied with flour. This is because the flour is measured in pounds, whereas the sugar and margarine are given in ounces. Similarly, there seem to be half as many eggs (3) as there are margarines (6)! Hmmm! What is going on here?

Graphs C3 and C4 compound all of the above problems with some additional foolishness along the vertical axis. C3 has contrived to have the vertical scale the wrong way round which has resulted in the largest data values having the shortest bars, and vice versa. The vertical scale in graph C4 has been drawn with unequal intervals which means that you can't trust the height of each bar to give an honest impression of the value it is meant to represent.

Verdict
Forget it!

Confusing percentages

> .... *If the standard of driving on motorways justified it, then I think a case for raising the speed limit would exist. As it is, surveys have shown that over 85% of drivers consider themselves to be above average! In reality, the reverse is more likely to be true.*
>
> From a letter to *Motorcycle Rider*, Number 13

As was indicated in Chapter 5, percentages are extremely useful for making fair comparisons between two numbers or two sets of numbers. For example, Table 6.4 shows the figures for church membership in the United Kingdom for over a fifteen-year period coming up to the end of

	BASE YEAR	15 YEARS LATER
Trinitarian churches	8.06	6.77
Non-Trinitarian churches	0.33	0.46
Other religions	0.81	1.86

Table 6.4   Adult church membership in the United Kingdom (millions)
Source: Adapted from *Social Trends 22*

the millennium. The data have been presented in three categories: Trinitarian churches (including Anglican, Presbyterian, Methodist, Baptist and Roman Catholic), Non-Trinitarian churches (including Mormons, Jehovah's Witnesses and Spiritualists) and other religions (including Muslims, Sikhs, Hindus and Jews).

Because these figures differ so widely, it is hard to make direct comparisons between the different religions and say how they have declined or grown. In this sort of situation, calculating the percentage changes is a useful way of comparing the *relative* changes for each religious group. Thus, relative change of membership for:

Trinitarian churches $= \dfrac{6.77 - 8.06}{8.06} = \dfrac{-1.29}{8.06} = 0.160$ or -16.0%

non-Trinitarian churches $= \dfrac{0.46 - 0.33}{0.33} = \dfrac{0.13}{0.33} = 0.394$ or 39.4%

other religions $= \dfrac{1.86 - 0.81}{0.81} = \dfrac{1.05}{0.81} = 1.296$ or 129.6%

Using these percentage changes we can now compare the rates of growth or decline directly over the fifteen-year period. Roughly speaking, the Trinitarian churches have lost around one in six of their flock while the non-Trinitarian churches have gained about two members for every five that they had at the start of the period. The other religions have grown more dramatically, with their numbers having more than doubled.

'A welcome slowdown' was how, in the 1990s, John Patten, then the UK Home Office minister described the 16 per cent rise in crime figures one year which was less than in previous years. This tactic is very much the stock in trade of politicians when asked to respond to an unpalatable percentage, namely to ignore the size of the change and concentrate only on whether the rate of change is going up or going down. Note that this 'slowdown' still represents a one in six increase over the previous year so reported crime rates were still rising rapidly. To claim that the *rate* of increase is not quite as fast as it was before is a little disingenuous – a bit like saying that we lost the match 10–0 – but at least we improved on the 12–0 drubbing we sustained the week before! The same gambit is used over the rate of inflation. When a politician claims that the rate of inflation has fallen, many people mistakenly believe that this implies that *prices* must therefore have fallen. On the contrary, any positive rate of inflation, whether it happens to be rising or falling, reflects rising prices. For the rate of inflation to fall from, say 8% to 6% simply means that prices are continuing to rise but at a slightly slower rate than before.

Staying in the world of politics, here is an example where a percentage was not used but perhaps should have been. The theme is poverty.

What does it mean to be 'in poverty'? Clearly, one person's notion of poverty will not be the same as another's. To some extent, our view of poverty is relative to the level of wealth around us – a poor family in the USA could be seen as being very well-off in, say, Ethiopia during a major famine.

Governments and other organizations such as the Low Pay Unit are interested in analysing the patterns in poverty so that the groups in greatest need can be clearly identified. An essential element in any such investigation is to have agreement between government and pressure groups on some unambiguous criterion for measuring what we mean by 'low pay'. An agreed measure of 'low pay' in the UK is earnings which are less than 50% of average (median) income. The figures in Table 6.5 indicate the relative numbers of children in poverty in the UK, by status of household.

STATUS OF HOUSEHOLD	NUMBER OF CHILDREN
Parent in full-time work	1025
Lone parent	938
Unemployed parent	763
Pensioner	320

Table 6.5 Typical numbers of dependent children in households below 50% of average income, analysed by family type
Source: Estimated from national UK data

EXERCISE 6.1. What's that as a percentage?

Convert the figures from Table 6.5 to percentages. What do they reveal about which of the four categories of household contains the greatest number of people in poverty?

Comments below

The total number of children here is

1025 + 938 + 763 + 320 = 3046

The percentage figures are calculated as follows:

Status of household		% children
Parent in full-time work	$\dfrac{1025}{3046} \times 100$	= 34
Lone parent	$\dfrac{938}{3046} \times 100$	= 31
Unemployed parent	$\dfrac{763}{3046} \times 100$	= 25
Pensioner	$\dfrac{320}{3046} \times 100$	= 10

Representing these percentages on a bar chart produces the graph in Figure 6.1.

The message of these figures seems to be that the largest number of children in poverty are from households where there is a parent in full-time work. However, although this conclusion is not wrong, the percentages which we have calculated do not tell the whole story and are therefore rather misleading.

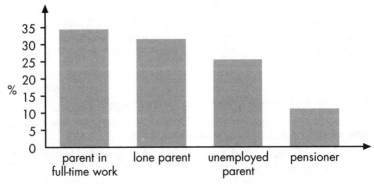

Figure 6.1 Bar chart showing types of household in greatest poverty

STATUS OF HOUSEHOLD	NUMBER OF CHILDREN IN POVERTY	TOTAL NUMBER OF CHILDREN
Parent in full-time work	1025	9330
Lone parent	938	1250
Unemployed parent	763	930
Pensioner	320	470
Total		11 980

Table 6.6   Number of children below 50% of average income, analysed by family type
Source: Estimated from national UK data

What is missing from this analysis is the total numbers of children in each of these four categories. These data have been added in Table 6.6.

As can be seen from the last column in this table, there are many more children from other households in the category 'parent in full-time work' than from any other. It is therefore not surprising that this category of household produces the largest number of children in poverty. What is a more useful calculation in this context is the percentage of each separate group of households in poverty. For example,

percentage of children in poverty from lone parent households

$$= \frac{938}{1250} \times 100 = 75\%.$$

EXERCISE 6.2. Recalculating the percentages

Calculate the other three percentages in this way and draw the results as a bar chart. How do you interpret these figures?

Comments below

When the percentages are calculated in this new way, based on the total number of children in each type of household (see Table 6.7 and Figure 6.2), it is clear that the majority of children from lone parent, unemployed and pension headed households are in poverty while only about one in ten children from full-time worker headed households is similarly economically disadvantaged.

STATUS OF HOUSEHOLD	% CHILDREN
Parent in full-time work	$\frac{1025}{9330} \times 100 = 11$
Lone parent	$\frac{938}{1250} \times 100 = 75$
Unemployed parent	$\frac{763}{930} \times 100 = 82$
Pensioner	$\frac{320}{470} \times 100 = 68$

Table 6.7   Percentage of children in poverty based on size of each household status category

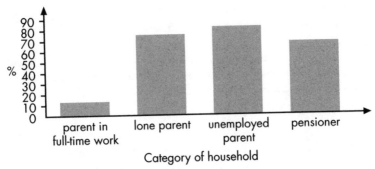

Figure 6.2   Bar chart showing percentages in Table 6.7

Inappropriate averages

Overheard in a bar in Glasgow:

When a Scot emigrates to England it raises the average level of intelligence in both countries.

We shall start with a situation where calculating an average might have been helpful but it wasn't actually used. The following cutting comes from an article on the British car industry.

Ford's job losses will reduce its manual work force to 27 000. This year it plans to make 450 000 cars. But Nissan's factory in

Sunderland will be producing 270 000 cars a year by next year with a work force of just 4500. Toyota's annual target is 200 000 cars with 3300 workers.

EXERCISE 6.3. Interpreting a jumble of figures

What do these figures appear to say about the relative productivity of Ford, Nissan and Toyota?

Comments below

When contained within a piece of text like this, the six numbers mentioned are rather difficult to interpret. It may clarify things to present the numerical information as shown in Table 6.8.

	CARS	WORK FORCE
Ford	450 000	27 000
Nissan	270 000	4 500
Toyota	200 000	3 300

Table 6.8   A neater summary of the data

The whole thrust of the article was that the Ford car-making process was less efficient in its use of workers than that of the two Japanese firms. A simple average would make this point rather more clearly, as illustrated in Table 6.9.

	CARS	WORK FORCE	AVERAGE NUMBER OF CARS PER WORKER
Ford	450 000	27 000	$\dfrac{450\,000}{27\,000} = 16.67$
Nissan	270 000	4500	$\dfrac{270\,000}{4500} = 60$
Toyota	200 000	3300	$\dfrac{200\,000}{3300} = 60.6$

Table 6.9   Including the averages

For another example of the way that averages can affect the way we make sense of information, let us return to the late 1980s, when Margaret Thatcher, the then prime minister of the UK, stated that:

Everyone in the nation has benefited from the increased prosperity – everyone.

Her remarks point to a common error with averages, which is to believe that they provide information about the *spread* of values, or in this case about the distribution of income and wealth. In fact, averages provide no information about the spread of values within a data set. Indeed, as the following simple example shows, it is possible for the average earnings to rise while the majority of people are actually worse off.

Imagine a small firm with 5 employees, all of whom were earning, say, £100 per week. The following year, the salaries were regraded. Four of the employees had their salaries reduced to £90, while the lucky fifth got a rise to £180.

EXERCISE 6.4. Far better or far worse?

What effect did this have on the average earnings of the five employees?

Comments below

Average earnings before the regrading

$$= \frac{100 + 100 + 100 + 100 + 100}{5} = £100$$

Average earnings after the regrading

$$= \frac{90 + 90 + 90 + 90 + 180}{5} = £108$$

So, the average earnings have increased (by 8%) but actually four out of the five people are less well off than before.

This is clearly a highly artificial example but in principle it describes what happened in the UK during the 1980s under Mrs Thatcher's government. Although average real earnings (i.e. earnings taking account of the effects of inflation) rose during this decade by something of the order of 30%, the widening inequalities resulted in many more people falling into debt and poverty. For example, have a look at the figures in Table 6.10.

	APRIL 1980		APRIL 1990	
	Men	Women	Men	Women
Highest decile	183	116	467	317
Upper quartile	143	91	347	245
Median	113	72	258	177
Lower quartile	90	58	193	136
Lowest decile	74	49	150	111

Table 6.10 The distribution of the gross weekly earnings (rounded to the nearest £) of men and women in April 1980 and April 1990

How, then, can we compare the spread of earnings for these four columns of data? The simplest measure of spread to use here is the inter-quartile range. For example, taking the earnings column for women in April 1980:

$$\text{Inter-quartile range} = £91 - £58 = £33$$

EXERCISE 6.5. Calculating the inter-quartile ranges

Calculate the inter-quartile ranges for the other three columns of data. Use your answers to compare the spreads of earnings for men and women over the decade between 1980 and 1990.

Comments on page 108

Clearly the inter-quartile ranges for 1990 (£154 for men and £109 for women) are much wider than for 1980 (£53 for men and £33 for women). However, since the numbers in the 1990 columns are larger than in the 1980 columns, it isn't very surprising that the inter-quartile ranges for 1990 should work out to be more than for 1980. You may remember that this issue cropped up in the previous chapter. There it was suggested that the only way a fair comparison can be made between two measures of spread is if they are *standardized*. The usual way to standardize the inter-quartile range is to divide it by the median. For example, the calculation for 1980 women is as follows:

$$\text{Standardized inter-quartile range for 1980 women} = \frac{33}{72} = 0.46$$

> **EXERCISE 6.6. Standardizing the inter-quartile range**
> **Calculate the standardized inter-quartile range for the other three columns of data. Use your answers to compare the spreads of earnings for men and women over the decade between 1980 and 1990.**
>
> *Comments below*

On the basis of the standardized inter-quartile ranges calculated in Exercise 6.6, there is now some objective evidence of how earnings inequalities widened over the 1980s. The standardized inter-quartile ranges for both men and women in 1980 were just under 0.5 and by 1990 they had risen to around 0.6.

Summary

This chapter provided some examples of ways in which statistics can mislead. It covered three key areas where statistical jiggery pokery tends to flourish – graphs, percentages and averages. The chapter ended with a longer case study where a suitable measure of spread, rather than a simple average, was needed in order to come to a sensible conclusion.

Comments on exercises

Exercise 6.5

For 1980 men, inter-quartile range = £143 − £90 = £53

For 1990 men, inter-quartile range = £347 − £193 = £154

For 1990 women, inter-quartile range = £245 − £136 = £109

Exercise 6.6

For 1980 men, standardized inter-quartile range $\dfrac{53}{113} = 0.47$

For 1990 men, standardized inter-quartile range $\dfrac{154}{258} = 0.60$

For 1990 women, standardized inter-quartile range $\dfrac{109}{177} = 0.62$

7 | CHOOSING A SAMPLE

You will need a calculator for this chapter.

Keith Parchment was arrested in April 1986 on suspicion of robbery. A major dispute at his trial concerned whether he had made a true confession or whether, as he claimed, it had been falsified by the police. In court, the detectives stated that Mr Parchment's original confession had taken them 31 minutes to write down. This worked out at a rate of 174 characters per minute. In order to test this out, his statement was written out again by one of the investigating team. This experiment revealed that rewriting his statement took 44 minutes to complete. So, was the version provided by the police a forgery?

The key question here was just how likely it might be that someone could (or would choose to) write legibly without a break at a rate of 174 characters per minute for 31 minutes. On the basis of the one person tested, this does seem unlikely. However, perhaps the person tested was a particularly slow writer. Simply choosing a sample of one person does not seem sufficient to come to a conclusion with any degree of confidence. Taking the population as a whole, how likely is it that someone chosen at random would have a writing speed of 174 characters per minute? We could always test the entire adult population of the UK, but this would prove far too costly and time-consuming.

In the event, a sample was taken of writing speeds among 17 staff at a forensic science laboratory. The participants were asked to try to ensure that their writing was as legible as that on the statement. The results of this experiment produced writing speeds which ranged from a minimum of 124 characters a minute to a maximum of 158 characters a minute. On the basis of this larger sample, Keith Parchment now had much firmer evidence that his evidence had indeed been falsified by the police and this was a key factor in his subsequent acquittal.

This example illustrates some important features of sampling which will be highlighted in this chapter. In general, we choose a sample in order to measure some property of the wider population from which it was taken. This may be in order to discover certain information about a single population (the extent of childhood illness, how many cigarettes people smoke, and so on), or the sampling may be part of an investigation into whether two or more populations are measurably different (are there regional differences in childhood illness, do boys smoke more than girls, and so on). A 'good' sample is one which fairly represents the population from which it was taken. This chapter looks at some of the key issues involved in choosing a 'good' sample, and starts with a definition of what we mean by the term 'sampling frame'.

The sampling frame

When talking about sampling in statistics, it is useful to separate out three related but distinct levels of terminology. Starting at the top level first, these are known as the 'population', the 'sampling frame' and 'sample'. Figure 7.1 below suggests that the sampling frame is (often) a subset of the population and the sample, in turn, is a subset of the sampling frame.

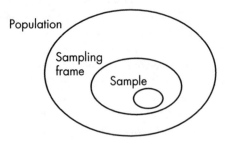

Figure 7.1 Population, sampling frame and sample

■ *Population* – in statistics, this term 'population' isn't restricted in meaning to refer just to a population of animals or humans but is used to describe any large group of things that we are trying to measure. This might be a precise measure of the length of each item of a particular manufactured component, or the weight of each bag of crisps coming off a production line, or the life in hours of a particular brand of light bulbs. In general, manufacturers need to monitor their production process to ensure that their grommets or their bags of crisps, etc are continuing to meet the required standard. Also, governments expect to collect detailed information about many aspects of the human populations for whom they provide services and from whom they collect money. It would be too expensive and time-consuming to measure each grommet or weigh each bag of crisps or interview each person in the population, so they all usually resort to some sort of sampling technique. The size of the population, i.e. the number of items in the population, is usually denoted by an upper case N.

■ *Sampling frame* – this is a list of the items from which the sample is to be chosen. Often it is the case that the sample is taken directly from the population, in which case the sampling frame would simply be a complete list cataloguing each item in the population. Sometimes, however, the population is simply too large and complicated for each element in it to be itemized accurately. For example, if the government wishes to estimate the national acreage of land growing oilseed rape, a suitable sampling frame would be a list of farms in the country. However, it is a major undertaking to list all the farms in the UK and it is likely that a number of small-holdings will be omitted.

■ *Sample* – this, in turn, is the representative subset of the sampling frame which is chosen as fairly as possible to represent the entire population. The term 'sample size', usually denoted by a lower case letter 'n', refers to the number of items in the sample. Where more than one sample is taken, it is worth stressing that 'n' is the number of items selected in a particular sample and not the number of samples taken. The (rare) situations where the entire population is 'sampled', i.e., where $n = N$, is known as a 'census'.

The tale of Keith Parchment illustrates two very important features of sampling. Firstly, if a sample is too small, it may not provide a very reliable or representative cross-section of the population as a whole and indeed may give a wrong impression of it. Secondly, there is the other side to this coin, namely that large samples, although they may be highly representative and accurately reflect the population, are usually also very expensive and time-consuming to carry out. In practical terms, therefore, sampling is always a compromise between accuracy on the one hand, and cost on the other.

Random sampling

There are several techniques for choosing a representative sample, of which probably the best known is *random sampling*. Its basic definition is as follows.

A random sample is one in which every item in the population is equally likely to be chosen.

With random sampling, there also needs to be some sort of random-generating process used to select each item. Suppose, for example, we wished to select a sample of size 50 (sample size, $n = 50$) from a population consisting of a sampling frame of size 1000 (i.e. $N = 1000$). The usual approach would be to allocate a number to each item in the sampling frame (from 1 to 1000 or perhaps from 000 to 999) and then find some way of randomly generating 50 numbers from within this range. The population items which had been allocated these numbers then become the 50 items which form our sample.

There are two common methods of random sampling – sampling with replacement and without replacement. Replacement means simply putting an item back into the sampling frame after it has been selected. Although replacement is sometimes thought of as being a more correct form of random sampling, in most practical sampling situations items are not replaced after they have been selected.

A useful mental image for a method of generating random numbers is a bingo 'random number generator' (RNG). Typically this is a large Perspex box containing 90 ping-pong balls numbered 1 to 90, which are tossed around by jets of air. The balls are released through a chute and displayed for all to see. The bingo caller then reads out the numbers in the time-honoured fashion – 'legs-eleven', 'blind-forty', 'all the sixes, clickety click, sixty-six', etc. This continues until a punter's card has been filled and the prize is awarded. The balls are then put back into the RNG and the process begins again with a new game.

> **EXERCISE 7.1. Sampling with and without replacement**
>   a) *As described above, is the bingo RNG an example of sampling with replacement or without replacement?*
>   b) *Why do you think the bingo RNG is designed in Perspex?*
>
> **Comments below**

Once a bingo number has been selected, it is no longer available for re-selection in the same game, so it is an example of sampling without replacement.

There may be several reasons why the box is made from Perspex. One may be that it is an interesting looking device which brings excitement and a sense of theatre to the game. But another important consideration is that the process is seen to be fair by all concerned. This is one reason why the ping-pong balls are usually set out on a rack for the punters to see once the balls have been selected. Clearly selections have the effect of producing winners and losers – of money, power, prestige, and so on – and throughout history such events have been traditionally associated with cheating. So, being fair, and being seen to be fair in the selection process, are both essential requirements of random sampling.

There have been many random generators over the centuries which have served a rather different purpose than to select a representative sample. For thousands of years, humankind has tried to divine the future by interpreting the outcomes of chance events. Of course, those who believe in this notion of 'chance divination' may describe the process rather differently. According to Marshall Cavendish in the book *Paths to Prediction* (Marshall Cavendish Books Ltd, 1991), it is a process in which we 'give Fate an opportunity to make its intentions known by using the operations of Chance'. The view of life implicit in this quotation appears to be one in which our Fate has been pre-ordained and we can tap its knowledge by means of any number of random generators – cards, tea leaves, bumps on the head, coins, dice and so on. For example, *I Ching*, or *The Book of Changes*, in Chinese, is an ancient philosophy which is founded on the centrality of Fate, and indeed Confucius turned to it for guidance and information. The randomization process which it involves can be carried out by using either three brass coins or a bunch of yarrow sticks. In the case of the coins, the 'heads' is given a value of 3 and the 'tails' has a value of 2. The three coins are tossed, giving a combined score of 6, 7, 8 or 9. This is carried out six times in all, thereby generating a six digit number, say, 687986. *The Book of Changes* matches this number up to a fragment of poetry which, hopefully, reveals its mysteries in the context of the question you have asked.

Clearly there is more to doctrines like *I Ching* than the mere one-off prediction of a future event. At the heart of the divinations lies a set of 'truths' and teachings which can provide inspiration and guidance for the whole of a person's life and for their entire community.

However, the other aspect of such predictions is the underlying belief that the chance events in question do not produce arbitrary or random outcomes but that they are somehow controlled and influenced by some unspecified supernatural force. This runs exactly counter to the statistician's view, which is that each outcome of a truly random event is indeed arbitrary, and the only underlying pattern is the obvious statistical one of equal likelihood – i.e. that, in the long run, you would expect each number to come up roughly the same number of times (not *exactly* the same number of times because you would expect some degree of variation).

Let us now return to random sampling with a simple example. Suppose that you have bought a ticket for a concert and find you are unable to go. In an unaccustomed burst of generosity, you decide to give the ticket to one of three close friends. But all are equally deserving and you owe no one of them a special favour. How do you choose which friend should receive the ticket?

Clearly this 'typical everyday' scenario is leading irrevocably to some sort of random selection procedure! One possibility might be to use the method of drawing straws and select them, in turn, on behalf of each friend. Whichever friend matched up with, say, the longest straw, would get the ticket. Another approach might be to toss a die. If the die shows a 1 or a 2, then friend A gets the ticket, if it shows a 3 or 4, then friend B gets it, and so on. But whichever approach you choose, two basic elements must be present, namely:

> (a) each friend must be matched up to a particular object (such as a straw) or number or set of numbers (as in the case of the die);

(b) there needs to be a fair mechanism for selecting the object or number such that each has an equal chance of selection.

EXERCISE 7.2. Some folks have all the luck

Here are two alternative methods of selection. Are they fair and, if not, then why not?

Method 1   Spin a coin between two people, A and B. If it shows 'heads', then A loses, otherwise B loses. Now spin a coin between the winner and person C. Whoever wins this spin gets the ticket.

Method 2   Toss two coins. If they show double-heads, A wins. If they show double-tails, B wins. If they show one head and one tail, then person C wins.

Comments below

Both of these methods fail on the grounds that the odds of winning are not equally shared among the three people. The unfairness of Method 1 lies in the fact that person C has got a 'bye' into the second round and has only one hurdle to overcome in order to win the ticket, whereas A and B both have two hurdles to overcome. C is actually twice as likely to win as either A or B. Method 2 also has a bias in favour of person C, but for a different reason. The explanation may be clearer if the four possible outcomes are written out. These are HH, HT, TH and TT. Each of these four outcomes is equally likely, so there is a one in four chance of any one of them occurring. However, note that A wins only when HH appears, B wins only when TT appears, but C has two chances to win – C will win with either HT or TH. Thus, once again C is twice as likely to win the ticket as either A or B.

Generating random numbers

Coins and dice are quite good random number generators in that their symmetrical shapes ensure fairly random sequences. However, for selecting samples of say, 15 or 20 or more from larger populations, the whole thing becomes a bit of a chore. You may be thinking that it would all be much easier if the throws of the die had already been done for you and

00	81668	75363	62126	65806	71928	30458	17405	39056	52083	20028
01	09065	19470	15770	43347	41754	63327	09071	56236	63510	45541
02	32519	12965	30543	88542	18830	31744	00980	43126	32154	32796
03	98740	98054	30195	09891	18453	79464	01156	95522	06884	55073
04	85022	58736	12138	35146	62085	36170	25433	80787	96496	40579
05	17778	03840	21636	56269	08149	19001	67367	13138	02400	89515
06	92409	79891	60979	67158	07958	72053	36272	78804	42477	40338
07	49487	52802	62058	87822	14704	18519	17889	45869	36752	54958

etc.

Figure 7.2 Random number tables (an extract)

the scores were set out in a table. In fact, this is exactly what a *random number table* provides, although the numbers are randomized in the range 0 to 9, rather than 1 to 6.

Figure 7.2 shows an extract from a random number table of the sort that can be found at the back of many statistics text books. Such tables hardly make for gripping reading, but one or two points are worth noting. Firstly, you can see how the numbers have been grouped in fives. This is simply to make it easier to read them. Also, unlike most sets of tables, there is no natural starting point – you can take any randomly chosen point in the table as the starting point and then simply read along the line from there, taking the numbers one at a time or two at a time etc, whichever is appropriate for your needs.

An alternative to using a random number table is to generate random numbers from a computer or calculator. Many scientific calculators possess a key marked something like 'Rand' or 'Ran #:' which, typically, will produce a random decimal number in the range 0 to 1. Alternatively it may be possible to use this facility to generate whole numbers randomly within a specified range.

Using random numbers to select a sample

Let us now look at how random number tables might be used in choosing a sample for a survey. To make it more interesting, the process is described in the context of an everyday example which might require the use of random number tables. We shall imagine that a hairdresser, called Marti, wishes to start a home hairdressing service in her area which she plans to call 'Home-cut'. Before she starts sculpting her first coiffure in the patrons' parlours, it would make sense to check out whether local people will be willing to part with their money for this service, and to establish what sort and price of service they might like. Marti decides that she is prepared to cycle up to a two mile radius of her home, so she can literally 'map out' her potential customers on a street map of the town. She decides to start with a fairly typical street, Woodcote Street, from within this area and sample the views of fifteen of the residents, asking them whether or not they would be prepared to use the service at least once a month. But she doesn't want to knock on people's doors to ask for an interview if she doesn't know their name. She also wishes to choose a representative sample of the target population.

Register of Electors

A useful publication for many kinds of sample survey work and one that is ideal for Marti's purposes is the local *Register of Electors*. For each electoral ward of each district, this register lists the private addresses and the names of all of the occupants of voting age in each residence. It can be bought by anyone for a fairly nominal sum and is available from local town halls and reference libraries. The registers are popular with banks, building societies and credit companies as a way of checking that new account-holders are in fact whom they say they are. They are also used by marketing agencies as a source of potential customers who live at the 'right' sort of addresses from a selling point of view. Each register is updated annually in the autumn by a small local staff, co-ordinated by their Electoral Registration Officer. Heads of households are required, by law, to complete a form stating the name and voting status of each member of the household.

Figure 7.3 shows the extract from the electoral register that refers to Woodcote Street. As you can see, the register is a cheap solution to the problem of having to avoid your neighbours because you have forgotten their name and, after all these years, it is now too embarrassing to ask. There are a number of other interesting things it can tell you about them as well. The information is organized in three columns, which show, respectively, the person's electoral number, their name and finally the number of their house. It is also interesting to see which houses share the same postcode – in this example the odd-numbered houses on one side of Woodcote Street have a different postcode to the houses on the other

WARWICK & LEAMINGTON Constituency					CV32 6TS	
Polling District DA–LEAMINGTON MANOR			1105	CORBETT, Mary-Lee.	2	
WOODCOTE STREET			1106	CORBETT, Thomas G.	2	
		CV32 6TZ	1107	HILL, Mary H.	4	
1064	BENNETT, Brian P.	1	1108	HOCKE, Ruth A.	6	
1065	BENNETT, Jane G.	1	1109	HOCKE, William R.	6	
1066	BENNETT, Michael F.	1	1110	DUFFY, Kevin T.	8	
1067	BENNETT, Anne S.	1	1111	CORLEY, Doris S.	10	
1068	BENNETT, Peter L.	1	1112	MASON, Gerald H.	12	
1069	TRASK, George L.	3	1113	MASON, Sarah J.	12	
1070	TRASK, Emma R.	3	1114	HUGHES, Mabel V.	14	
1071	GEORGE, Helen	5	1115	HUGHES, John M.	14	
1072	UNITT, James G.	7	1116	SMITH, Hilda E.	16	
1073	UNITT, Grace E.	7	1117	SMITH, Brian W.	16	
1074	2/01/01 UNITT, Jean L.	7	1118	SMITH, Edna M.	16	
1075	EVANS, Marie L.	9	1119	WADE, Iris G.	18	
1076	EVANS, Robert	9	1120	DUCKER, Herbert K.	20	
1077	WILSON, Annie L.	11	1121	DUCKER, Ethel M.	20	
1078	WILSON, Hazel E.	11	1122	DUCKER, Georgina J.	20	
1079	OAKLEY, Edith D.	13	1123	FRAYN, Shirley	22	
1080	OAKLEY, Winston H.	13	1124	HUNT, David R.	22	
1081	SONLEY, Mary	15	1125	HARRISON, Terence	24	
1082	WALLSGROVE, Carol G.	17	1126	HARRISON, Madelein	24	
1083	WALLSGROVE, Iain G.	17	1127	HARRISON, Mandy J.	24	
1084	JONES, Valerie E.	19	1128	DAY, Edna J.	26	
1085	JONES, David Q.	19	1129	DAY, Arthur F.	26	
1086	JONES, Douglas B.	19	1130	DAY, Paulyn J.	26	
1087	EDEY, Andrew H.	21	1131	19/07/01 DAY, Colin	26	
1088	EDEY, Carol D.	21	1132	HOYOW, Helen T.	28	
1089	WEBB, Elsie S.	23	1133	HOYOW, Jeffery.	28	
1090	JOHNSTON, Stuart D.	25	1134	HOYOW, Phyllis T.	28	
1091	McNAIR, Doris F.	25	1135	HOYOW, Hilary J.	28	
.1092	NEWBERY, Ian J.	27	1136	HOYOW, John E.	28	
1093	NEWBERY, Ida T.	27	1137	KAI, Paul.	30	
1094	NEWBERY, Dorothy T.	27	1138	McGREANOR, Alan T.	30	
1095	NEWBERY, Angus D.	27	1139	FU, Ahchun	32	
1096	MARSHALL, Stuart T.	29				
1097	MARSHALL, Margaret Y.	29				
1098	PETTS, Hepzibah	31				
1099	SIMS, Olive R.	31				
		CV32 6UH				
1100	METHERELL, Isabella T.	1, HOPE GARDENS				
1101	HARDING, Nancy R.	2, HOPE GARDENS				
1102	HARDING, Peter G.	2, HOPE GARDENS				
1103	HARDING, Anne T.	2, HOPE GARDENS				
1104	METHERELL, Ralph A.	3, HOPE GARDENS				

Figure 7.3 Extract from an electoral register (names have been changed)

side, while the three houses in Hope Gardens have a postcode all to themselves.

Now look at elector number 1074, whose name is preceded by a date. This signifies that Jean Unitt was not of voting age at the time the census was carried out but will be celebrating her eighteenth birthday on the date in question and would therefore be eligible to vote from then. (This is another useful piece of information for a caring neighbour who doesn't want to miss out on the thoughtful present.)

EXERCISE 7.3. Selecting fairly

Using Figure 7.3 as your sampling frame and the table of random numbers in Figure 7.2, think about how you would choose a fair sample of size 15 from the people in Woodcote Street.

Comments below

The polling numbers corresponding to these residents of Woodcote Street run from 1064 to 1139 inclusive, giving a sampling frame of size 76. So a sample of size 15 represents around one fifth of the sampling frame. We can ignore the first two digits of the polling numbers and concentrate on the last two, the 64, 65, and so on. Now we can turn to the random number table in Figure 7.2. Since we have two-digit numbers in the sampling frame, it is necessary to select the random numbers from Figure 7.2 in pairs. Starting at the beginning of any row, say row 03, this gives the following set of two-digit numbers:

98   74   09   80   54   30   19   50   98   91 and so on.

When you get to the end of row 03, simply continue on to the beginning of row 04, and so on. We are now ready to select the names of the fifteen people. Number 98 corresponds to polling number 1098 and this is Hepzibah Petts, who lives at number 31. You can now tick her name on the polling register as the first person to be selected for Marti's sample. However, perhaps you can foresee that we will have problems when we get to the fifth number, 54, because there is no-one in Woodcote Street with a polling number of 1054. Fortunately random numbers come cheap, so the easiest solution is to ignore any random number which doesn't occur within the sampling frame and simply go on to the next pair of digits. Also, you can see that the ninth random two-digit number,

98, is one which has already occurred, so again it will be ignored. Incidentally, this is why it is a good idea to tick each person's name on the list as you go along, since you can see at a glance whether or not they have already been selected. What this means, of course, is that it will be necessary to sample more than fifteen pairs of random digits in order to get a sample of size 15, so it is a good idea to keep a running tally of the number of items currently in the sample as you go along.

EXERCISE 7.4. Go for it

Now complete the sampling procedure and identify the fifteen people from Woodcote Street who will make up Marti's sample.

Comments below

Your solution should look like this.

RANDOM NUMBER	ELECTORAL NUMBER	NAME	RUNNING TOTAL
98	1098	Hepzibah Petts	1
74	1074	Jean Unitt	2
09	1109	William Hocke	3
80	1080	Winston Oakley	4
54	—		
30	1130	Paulyn Day	5
19	1119	Iris Wade	6
50	—		
98	1098	—	
91	1091	Doris McNair	7
18	1118	Edna Smith	8
45	—		
37	1137	Paul Kai	9
94	1094	Dorothy Newbery	10
64	1064	Brian Bennett	11
01	1101	Nancy Harding	12
15	1115	John Hughes	13
69	1069	George Trask	14
55	—		
22	1122	Georgina Ducker	15

Table 7.1   The fifteen people sampled

Sampling variation

Marti noticed that nine of the fifteen, i.e., 60% of the residents chosen were female. Clearly the gender of her potential customers is an important consideration, and Marti may want to know how typical her sample is in this respect. Normally it is quite difficult to check the characteristics of the sample against those of the population or sampling frame from which it was taken. After all, the reason that you chose to take a sample in the first place was that the population was too large to analyse in its entirety. However, in this particular (and indeed slightly contrived!) example, we can check out the characteristics of the sampling frame fairly easily. A quick check of the names in the entire sampling frame of Woodcote Street register of electors reveals that 44 of the 76 residents, or 57.9%, are female. This is quite close to our sample proportion of 60%, but, given the fairly small sample size of 15, one might expect quite a wide variation if several such samples are selected. The final section of this chapter on sampling error explains how sample variation can be measured and used to say something about the degree of confidence with which population estimates can be made from sampled data.

Marti also wanted to check whether her sample was representative of the houses on either side of the street. This could be important since the even-numbered houses are on the more affluent north side of the street, and she might expect these residents to be able to afford a more up-market service. A check reveals that, ignoring Nancy Harding from Hope Gardens, there were equal numbers of even- and odd-numbered addresses represented in the sample, which compares very closely with the sampling frame as a whole (35 addresses were even-numbered and 36 were odd-numbered). You could not expect that repeated samples would produce such a close match between sample and sampling frame, but most of the samples might be expected to contain somewhere between four and eleven of both types of house.

A further consideration which Marti might wish to take account of in her survey is the ages of all the residents chosen. Unfortunately this is one piece of information that the Register of Electors does not provide.

Systematic sampling

Although random sampling is an attractive means of choosing a sample, it is not always the most convenient method. For example if you wished to find out about consumers' brand preferences for various goods, it makes more sense to stand outside a few selected supermarkets and shops and ask people as they come out. Three further sampling techniques that are commonly used are *systematic*, *stratified* and *cluster* sampling. This section focuses particularly on systematic sampling.

Most street or home-based polls are based on some sort of systematic sampling method. For example, if the polling agency is aiming to carry out, say, a 10% poll of a street, the interviewer may be instructed to knock on every tenth door or to stop every tenth passer-by. This might have been a sensible method for Marti in her survey of Woodcote Street. She wished to select 15 from a total of 76 residents in the street, which is roughly 1 in 5. So she will need to use some random selection method to choose the first person's electoral number and then repeatedly add five to select the other 14 members of her sample. Tossing a die is a suitable method of starting the process and let us suppose that this gives the starting value of 2 (had it shown up as 6 she would have had to toss again). So the first person Marti will visit is the second elector on the list, Jane Bennett. Table 7.2 shows the full sample selected by this method.

An obvious advantage of systematic sampling is that it guarantees an even spread of representation across the sampling frame. So, as can be seen from the table above, the method has ensured that Marti has sampled equally the even- and odd-numbered houses. However, the

ELECTORAL NUMBER	NAME	HOUSE NUMBER
1065	BENNETT, Jane G.	1
1070	TRASK, Emma R.	3
1075	EVANS, Marie L.	9
1080	OAKLEY, Winston H.	13
1085	JONES, David Q.	19
1090	JOHNSTON, Stuart D.	25
1095	NEWBERY, Angus, D.	27
1100	METHERELL, Isabella T.	1, HOPE GARDENS
1105	CORBETT, Mary-Lee	2
1110	DUFFY, Kevin T.	8
1115	HUGHES, John M.	14
1120	DUCKER, Herbert K.	20
1125	HARRISON, Terence I.	24
1130	DAY, Paulyn J.	26
1135	HOYOW, Hilary J.	28

Table 7.2   Systematic sample of every fifth elector

method will provide no certainty that the sample is representative in terms of the distribution of age or gender of the people sampled (this sample has produced only 7/15, or 46.7%, females, which is a slightly lower than expected proportion of 57.9% in the sampling frame). It must be remembered that a sample size of 15 is very small indeed and randomly selecting such a small sample provides no guarantees at all that it will be very representative of the population from which it was selected.

So in situations where it may be essential that the sample is evenly spread across the sampling frame, systematic sampling would appear to be a better method. However there are certain dangers inherent in systematic sampling, as the next exercise should reveal.

EXERCISE 7.5. Systematic bias
What biases might the following sampling procedures pick up?

■ Estimating yearly sales of a shop over a ten year period by sampling sales figures systematically every 6 months.

■ Estimating a household's monthly expenditure by sampling their bills and payments on the first week of every month.

■ Estimating the absentee rates of a firm by sampling their absentee rates every 5 working days.

Comments below

All of these examples carry the dangers of picking up a cyclical pattern in the population data and repeating it again and again. For example, sampling sales every 6 months will pick up seasonal fluctuations. So, if the first month was, say June, then every second month sampled would be a December, in which case the untypical pre-Christmas rush would be picked up each year. Similarly, household expenditure patterns are never uniform over a month. Many people arrange that standing orders are paid out at the beginning of the month and they may hang on until the benefit cheque or wage packet arrives before taking on a major shopping expedition. Finally, if a survey on absenteeism is carried out on the same day each week, certain biases such as the 'Monday morning syndrome' or the 'Friday afternoon syndrome' may distort the findings.

These examples of systematic bias are fairly obvious and it wouldn't be difficult to set the repeating samples cycle to avoid them. However, the danger with systematic sampling is that you may pick up an underlying cycle in the population that you weren't aware of, and so introduce bias unknowingly. There is no easy remedy here except to use a bit of common sense and imagination.

To end this section, it is worth mentioning two further sampling techniques – *stratified* and *cluster* sampling. There are two obvious 'strata' or layers in the Woodcote Street data, namely the socio-economic distinction between the odd- and even-numbered houses. If Marti felt that it was essential that these two sub-groups were equally represented, but wished to use a random technique, she could partition her sample into these two strata and select randomly from each stratum. This is a widely used technique and it requires that the strata are easily identifiable in the population. For example, the designers of a survey on smoking in pregnancy may wish to identify three categories of pregnant women – non-smokers, light smokers and heavy smokers – and try to ensure that each of these strata is represented within their sample in the same proportions as they occur in the overall population of pregnant women.

Suppose that a sample is required for interview from a large population such as all the inhabitants of the UK. Clearly the people selected for the sample may be very widely spread geographically and the costs of travelling to each separate address would be huge. *Cluster sampling* is a way of reducing the time and cost involved by restricting the sample to a smaller number of geographical areas. The population in each of these areas is a cluster and the sub-samples from each cluster are combined to produce the final sample.

Human error and sampling error

> I reckon the light in my 'fridge' must stay on all the time. I've checked hundreds of times and every time I open the door to have a look it is still on!

One of the dilemmas for any researcher is that he or she never really knows the extent to which their research observations affect or disturb the objects or people being observed. As with the 'opening the ''fridge'' door' example above, all surveys involve opening some sort of door on people's lives. Observers participate in this process and can never fully know the extent to which their presence has affected the phenomena that are being presented to them. All they can do is attempt to monitor their findings against those collected in a different way and then try to account for any differences. Most of all, the observer needs to be sensitive to these questions and make every attempt to tread lightly.

There are two main types of error that surround sampling. The first is the sort introduced by a badly designed sampling procedure, and some examples of these have been given earlier in the chapter. This might be called 'human error'. Sometimes error is deliberately and unfairly introduced into the sampling process in order to distort the findings. Examples of this are the original polls which were used to compile the pop music charts. The first singles chart, known as the Record Hit Parade, was released in 1952 and was published by the *New Musical Express*. (Before that, the charts were compiled from sheet music sales.) In those early days, a small number of record shops were selected and sampled for their record sales over the previous week. The problem was that the sample size was small and, once the identity of a 'chart shop' became known, agents would descend with wads of fivers to buy particular records and thereby bump up their chart ratings. This problem was tackled by increasing the number of 'chart shops', and today the Gallup polling agency responsible for compiling the charts samples from around 1000, or roughly one in four, of all record outlets.

Phone-in polls are a popular means by which some radio stations compile record charts, but these are even more open to distortion than sampling sales from record outlets. In 1986, when Capital Radio decided to compile the all-time Top 100 records, based on a listeners' phone-in poll, five Bros records featured in the chart. This probably says something about the promotional zeal of the listening 'Brosettes' and the fact that a phone call or ten costs less than the price of a single, particularly if mummy and daddy are footing the telephone bill.

Let us now turn to the second form of error. Due to natural variation, all sampling is inherently prone to error and this variation is an inevitable part of the random sampling process. It is this type of error which is referred to as 'sampling error'. Although sampling error cannot be eliminated, it is possible to quantify how much variation to expect under different conditions, and this allows us to gain some measure of the confidence with which we can make predictions about the population. For example, rather than using her sample to predict that, say, 185 people in her target population might be interested in a home-cut, Marti can use a measure of her sampling error to come up with a more useful statement such as that she is 95% confident that between 160 and 210 of the people in her target population might be interested in a home-cut. This 160–210 interval can be described as a *95% confidence interval* and this sort of statement is known as an *interval estimate*. Such a statement is particularly useful as a way of thinking through the worst-case and the best-case scenarios.

EXERCISE 7.6. Spot the connection

The width of the *confidence interval* is related to the *sample size*. How would you expect these two factors to be related?

Comments below

The larger the sample you take, the greater your confidence will be that the sample result is accurate, so therefore the narrower will be your confidence interval. This may appear to be slightly counter-intuitive (the notion of high confidence being linked to 'narrow' confidence) but it may be helpful to think of the confidence interval as a 'tolerance' – a low error tolerance is more comfortably associated with a high degree of confidence in the population estimate. The best way to be convinced about this relationship between confidence and sample size is to do a simple sampling experiment based on samples of different size. If you then plot the sample means you will find that the means which were calculated from the smaller samples are spread out more widely than the means taken from large samples. This basic idea is first illustrated with a simple example using dice. The same point is then made using a larger sample, this time with random numbers.

This first experiment is about finding the mean score when a die is tossed more than once. Regardless of how many times the die is tossed, the mean score that you would expect to get is around 3.5. (This is because the six outcomes 1, 2, 3, 4, 5 and 6 are equally likely to occur and the mean of these six numbers is 3.5.) Of course the means will not always equal 3.5 – sometimes they will be more and sometimes less. Now imagine two dice, one black and the other white. The black die is tossed twice and the mean of the two scores taken. The white die is tossed six times and the mean of the six scores is taken. In which case might you expect the mean to be closer to 3.5? Will it always be so?

EXERCISE 7.7. Dice games

Find a die and carry out a similar experiment to the one described below – this will only take you a few minutes and should ensure that you grasp the essential point of it. Check that the same outcome emerges as the one described here.

a) *Toss the black die twice and calculate the mean: e.g., 5, 1 → mean = 3.*
b) *Toss the white die six times and calculate the mean: e.g. 6, 2, 4, 3, 3, 1 → mean = 3.17.*
c) *Note which of the two answers is closer to 3.5.*
d) *Repeat the experiment several times. Overall, which die, the black or the white, tends to 'win' – namely to lie closer to 3.5?*

Comments below

You probably found that the white die, based on a mean from a larger sample size, tended to win this particular dice game. This result appears to confirm the earlier claim, that 'the larger the sample you take, the narrower will be your confidence interval'. However, this was a very crude and small-scale experiment. In Table 7.3, the same basic exercise has been done, but on a rather larger scale using random number tables.

The calculations in Table 7.3 were performed on a spreadsheet, which took a lot of the hard graft out of the activity! The column headed 'Random X' contains the first 40 random digits in row 07 of Figure 7.1. The first figure in column 2 shows the mean of the first two of these random numbers (i.e., 6.5 is the mean of 4 and 9). The second figure in column 2 shows the mean of the second and third random number, and so on. The figures in the other columns show the result of averaging the random numbers three at a time, five at a time and ten at a time, respectively.

RANDOM x	MEANS ($n = 2$)	MEANS ($n = 3$)	MEANS ($n = 5$)	MEANS ($n = 10$)
4				
9	6.5			
4	6.5	5.7		
8	6.0	7.0		
7	7.5	6.3	6.4	
5	6.0	6.7	6.6	
2	3.5	4.7	5.2	
8	5.0	5.0	6.0	
0	4.0	3.3	4.4	
2	1.0	3.3	3.4	4.9
6	4.0	2.7	3.6	5.1
2	4.0	3.3	3.6	4.4
0	1.0	2.7	2.0	4.0
5	2.5	2.3	3.0	3.7
8	6.5	4.3	4.2	3.8
8	8.0	7.0	4.6	4.1
7	7.5	7.7	5.6	4.6
8	7.5	7.7	7.2	4.6
2	5.0	5.7	6.6	4.8
2	2.0	4.0	5.4	4.8
1	1.5	1.7	4.0	4.3
4	2.5	2.3	3.4	4.5
7	5.5	4.0	3.2	5.2
0	3.5	3.7	2.8	4.7
4	2.0	3.7	3.2	4.3
1	2.5	1.7	3.2	3.6
8	4.5	4.3	4.0	3.7
5	6.5	4.7	3.6	3.4
1	3.0	4.7	3.8	3.3
9	5.0	5.0	4.8	4.0
1	5.0	3.7	4.8	4.0
7	4.0	5.7	4.6	4.3
8	7.5	5.3	5.2	4.4
8	8.0	7.7	6.6	5.2
9	8.5	8.3	6.6	5.7
4	6.5	7.0	7.2	6.0
5	4.5	6.0	6.8	5.7
8	6.5	5.7	6.8	6.0
6	7.0	6.3	6.4	6.5
9	7.5	7.7	6.4	6.5
Max 9	8.5	8.3	7.2	6.5
Min 0	1.0	1.7	2.0	3.3
Range 9	7.5	6.6	5.2	3.2

Table 7.3   Investigating sample means using different sample sizes

At the bottom of the table are given the minimum value, maximum value and range of each set of sample means. (Note that Range = Max − Min.)

Notice how the mean values in column 2 vary enormously from a minimum of 1.0 to a maximum of 8.5, giving a range of 7.5. However, as the sample size is increased to 3, then 5 and finally 10, this spread of the mean values is progressively reduced (to a range of 6.6, 5.2 and 3.2, respectively). Once again, this confirms the 'central finding', namely that means calculated from samples with larger sample sizes tend to lie closer to the true mean than means calculated from samples with smaller sample sizes. This is an important result in statistics and is the central plank in what is known as the 'central limit theorem'. In statistical terms, it is the key element in the Keith Parchment story that started this chapter off, namely that confidence about the results of sampling are increased as sample size is increased.

Summary

This chapter looked at a number of key ideas in sampling. The terms population, sampling frame and sample were distinguished and you were introduced to the idea of random sampling. These ideas were given a practical context, using an example based on the electoral register and a set of random number tables. You then looked at sampling variation and went on to consider several alternative sampling techniques – systematic, stratified and cluster sampling. Finally the distinction was drawn between human error and sampling error, and the chapter ended with an experiment designed to illustrate a key element in the 'central limit theorem', namely that means calculated from samples with larger sample sizes tend to lie closer to the true mean than means calculated from samples with smaller sample sizes.

8 COLLECTING INFORMATION

Data, give me data Watson! I can't make bricks without straw!
Sherlock Holmes

In Chapter 1, four key stages of a statistical investigation, known as the 'PCAI' cycle, were listed as follows:

- ■ P   Stage 1   pose a question
- ■ C   Stage 2   collect relevant data
- ■ A   Stage 3   analyse the data
- ■ I   Stage 4   interpret the results.

We now turn to the 'C' stage of the PCAI cycle, collecting relevant information. In any statistical work, what makes data relevant is determined by whether or not the information will help you to answer your central question. It cannot be stressed enough that the more purposeful and clearly formulated is your central question (stage 'P'), the easier it will be to make sensible decisions at the 'C', 'A' and 'I' stages later on.

The first guiding principle at the 'C' stage is to remember that data generation costs time and money, so avoid re-inventing the wheel. In other words, before you rush out and try to generate your own data, make sure that someone else hasn't already done it for you. Data which have already been collected by someone else and are awaiting your attention are known as *secondary source* data or sometimes simply as *secondary* data. In the absence of suitable secondary source data, you may have to carry out an experiment or survey yourself, in which case this would involve you in generating *primary source* data (also known as *primary* data). Libraries are full of valuable secondary source data and this is probably the first place you should look.

Secondary source data

Historically, the term 'statistics' derives from 'state arithmetic'. For thousands of years, rulers and governments have felt the need to measure and monitor the doings of their citizens. From raising revenue in the form of taxes to counting heads and weapons in times of war, the state has always been the major driving force in collecting and publishing statistical information of all kinds. And, as you will see from this section, they are still at it! Just a few of the most useful secondary sources of data are described here, most of which are from government sources.

Let us assume that you have identified a particular question for investigation. As has already been suggested, the first task is to check whether any suitable secondary source data are available. You enter your local library, but where do you look? A good starting place is usually the latest edition of *Social Trends*. This annual publication of the Government Statistical Service is prepared by the Central Statistical Office (CSO) and published by HMSO. As well as *Social Trends*, the best known statistical publications from the CSO are *Economic Trends* and *The Monthly Digest*. They also produce monthly Business Monitors and CSO Bulletins and have a wide range of computerized data on disk.

Social Trends

This excellent publication is crammed with interesting tables and charts about a wide variety of social and economic issues. At the time of writing, the data it provides are presented in the following 13 chapters.

1 Population
2 Households and Families
3 Education
4 Employment
5 Income and Wealth
6 Expenditure and Resources
7 Health and Personal Social Services
8 Housing
9 Environment
10 Leisure
11 Participation
12 Law Enforcement
13 Transport.

Depending on the particular original source being quoted, the data are based on UK, Great Britain or England & Wales but comparisons are given over time (anything from, say, 10 to 30 years) and against selected other countries (particularly EU countries). Each chapter ends with a bibliography listing the main sources used. These references are themselves useful should you not find the relevant item of secondary source data in the *Social Trends* publication directly. The appendix to *Social Trends* lists the major surveys used, and some of the best known of these are listed below.

The *Family Expenditure Survey* (FES), which investigates, annually, the income and spending patterns of a sample of around 7000 households in the UK.

The *Survey of Retail Prices*. This is carried out monthly by the Department of Employment and provides about 150 000 prices used in calculating the Retail Prices Index (RPI). (*The Employment Gazette*, HMSO, provides a useful, detailed and up-to-date summary of these data each month.)

The *New Earnings Survey* (NES), which collects information on an annual basis on the earnings of about 180 000 people. (This information is published annually in the *New Earnings Review*.)

The *British Social Attitudes Survey*. This annual survey solicits the opinions and attitudes of between three and four thousand individual adults each year, and the results are written up in the annual report, *British Social Attitudes*.

The *General Household Survey* collects a wide range of information on households from over 10 000 addresses, resulting in an annual publication of the same name.

The *Census of Population* takes place every 10 years (in 1991, 2001, etc) and is based on a restricted amount of information about personal details and the nature of households of all citizens of the UK. It is used as a source of reliable statistics for small groups and areas throughout the country so that government and others can plan housing, health, transport, education and other services. The 1991 census data were collected from some 24 million households. Local area statistics from the census were published separately for each county and all the key results of each county are also available in machine readable format.

One frustration with secondary data is that they are rarely in exactly the form that you want. You may think that you have collected information on gross earnings in the UK only to read the small print and discover that the figures refer to Great Britain only or that part-time workers have been excluded and you need the inclusive figure for purposes of comparison. Fortunately there is a simple remedy for such problems – it is called 'hard graft'! You need to be prepared to put in the time and effort checking out all the small print of any secondary data that you propose to use and making sure that they really are the figures that you want. As often as not you will need to chase up some additional data from elsewhere.

Data accuracy

All data that have been collected by measurement of some sort, whether secondary source or primary source, are inaccurate. This is so for a variety of reasons. Firstly, no measuring tool is perfectly accurate. If the information has derived from physical measurement, the ruler or weighing scales or thermometer, etc, can only be read off to a certain degree of accuracy. The following exercise makes the point.

EXERCISE 8.1. Paper thin

A block of 60 sheets of writing paper is measured at $\frac{1}{2}$ cm in thickness. How thick is each sheet?

Comments below

Now $\frac{1}{2}$ cm = 5 mm. So, the appropriate calculation seems to be:

Thickness of 60 sheets = 5 mm
Thickness of one sheet = $5 \div 60 = 0.083\ 333\ 3$ mm

However, how confident are you of this result to eight figures of accuracy? Probably not at all! There is no clear indication in the question as to the measuring method used and the degree of accuracy of the original measurement of $\frac{1}{2}$ cm. A difficulty with expressing numbers as fractions like this is that it is impossible to apply the normal conventions of accuracy. If the original measurement had been expressed as a decimal number (either 0.5 cm or 5 mm) it would be reasonable to assume an accuracy of 1 mm. In other words, the block of paper was between 4.5 and 5.5 mm in thickness. (Note that if the measurement had been carried out to an accuracy of, say, 0.1 mm, the result would have been written as either 0.50 cm or 5.0 mm.) If we take these two maximum and minimum values separately, we get the following range of values for the thickness of a single sheet.

Minimum thickness of one sheet $= 4.5 \div 60 = 0.075$ mm
Maximum thickness of one sheet $= 5.5 \div 60 = 0.0916667$ mm

An awareness of the range of possible values – from smallest to greatest – emphasizes the nonsense of quoting the original answer to 8 figures. In this and similar examples, it is sensible to *round* the answer to an appropriate degree of accuracy. Here an answer of 'about 0.08 mm' or even 'about 0.1 mm' would seem to be an appropriate estimate for the thickness of a single sheet.

A second source of data inaccuracy is human error. This may be due to the researcher misreading the instrument or perhaps miscoding or miscounting the category responses to a survey. An additional dimension to the incidence of 'human error' in survey work is that respondents may, either knowingly or unknowingly, be providing false information. There have been several well-documented historical examples of this in the field of anthropology, where European researchers have visited isolated 'primitive' island communities and reported complex and bizarre sexual rituals, only for subsequent researchers to discover that the 'primitive islanders' were, in fact, engaging in a jolly leg-pull exercise and had been making up whoppers for their own amusement!

Sometimes it is possible to identify and even quantify respondent error in follow-up surveys. For example, in the United States, the Bureau of the Census has a policy of resurveying a sample of the original respondents for verification. One of the questions asked of unemployed respondents is to state the number of weeks that they had been unemployed. In the follow-up interview, say five weeks later, one might expect those respondents who are still unemployed to produce an answer to this same question that was five weeks longer than their earlier answer. Some hope! In the event, only about one quarter give consistent responses. Of the other three quarters, some 'gain' weeks of unemployment between interviews and others 'lose' weeks of unemployment. The problem for the researcher is deciding which (if any) of these responses to believe. It is also worth noting that this example reveals a high likelihood of widespread inaccuracy in the responses from most respondents to most questions. Let's not fall into the trap of assuming that just because someone tells you something in an 'interview' situation that they really know the answer, or that their reply is necessarily true!

Primary source data

It may be that your research question is concerned with an issue that is too specific or too local for there to be suitable secondary source data available, in which case you may decide to collect your own data. This may require accurate measurement that involves carrying out some sort of scientific experiment, or perhaps designing a questionnaire and conducting a sample survey. In this section we will look at two of the key issues in this area – experimental design, and questionnaire design.

Experimental design

In 1990, the parents of two teenagers, both of whom had committed suicide, sued the rock group Judas Priest and their record company, CBS records. The parents alleged that the deaths were partly due to subliminal messages (the words 'Do it') concealed in one of the group's 1978 albums, Stained Class. In the event, the group and their record company were cleared of the charges by a Los Angeles court, but the case raised considerable interest in the whole area of the effects of subliminal messages, both in relation to advertising and for educational purposes. A number of commercial self-help audio tapes have been produced containing subliminal messages against a background of relaxing music, all designed to help people lose weight, or give up smoking or even to improve their memory. Over a five week period, in 1991, a group of people listened every day to a tape that contained subliminal suggestions for improving memory. Half felt that their memory had improved. Quite impressive, you may feel! However, the researchers also asked a second group to listen to a tape that they *thought* contained messages but in fact did not. Well, guess what? Roughly half of the second group reported the same thing. So what is going on here? According to the report of the British Psychological Society (*Subliminal messages in recorded auditory tapes and other unconscious learning phenomena*, 1992), the explanation may be as follows:

When someone expects something to be helpful, there can be a positive outcome, even if the product itself is useless. People tend to avoid the embarrassment of confessing to themselves that a financial investment has been wasted.

EXERCISE 8.2. Cure or con?

Suppose that a drug company has claimed to have invented a cure for the common cold. They have tried it out on ten cold-suffering volunteers and, indeed, all ten got better. Why might you not be convinced by their conclusions that the drug really works? How could you improve on the experimental design in order to give the drug a fairer test?

Comments below

Rest assured that drug companies would never get away with this experiment as there are strict rules to be observed in the manufacture and testing of drugs before they are licensed and made available to the general public. Here are some of the experimental weaknesses in the example above.

- Who says they are better? The states of being 'well' or 'ill' or 'better' or 'worse' are often highly subjective. Who is making the health judgement on these subjects and do they have a vested interest in the outcome? It is possible that the drug has simply suppressed the symptoms (headache, sneezing, watering eyes, etc) but the virus is not eliminated.

- Sample size – a sample of only ten subjects is much too small.

- Time scale – we don't know if recovery took place over a period of hours, days or years.

- Controls – there is nothing to compare with here. A key question is whether the subjects would have recovered anyway without this treatment.

- The *placebo* effect – this is sometimes also known as the 'feel good' factor. Particularly in the area of health, people often respond positively to psychological influences – simply as a result of swallowing a pill or drinking some nasty medicine. So the psychological benefits of being given treatment may be obscuring the chemical merits or demerits of the drug.

Proper drug testing is based on an experimental design which attempts to meet these objections.

On the question of 'who says they are better?', there need to be, where possible, objective measurable criteria (blood or urine tests, body temperature checks, and so on) to assess whether or not the subject's health has indeed improved after taking the drug.

A sample size of several hundred or several thousand would be required in order to test a drug rigorously. This is necessary for two reasons. Firstly, a reasonably large sample size is needed in order to carry out subsequent analysis of tests of significance at the level of significance required. Secondly, when licensed, the drug will not be administered to a homogeneous collection of patients, but rather to people of a wide variety of ages, body weights, medical backgrounds, and so on. It may also be offered to pregnant women. Thus, testing needs to be on a scale that is sufficient to establish with some degree of certainty whether or not the drug is safe for this range of potential users. Testing may also be aimed at establishing effective and safe dosage levels for adults and children separately.

The time scale over which the drug appeared to be effective also needs to be determined. This is important both to check that the drug really is working (people tend to recover on their own eventually if left untreated) and also to fine-tune the period over which the drug needs to be taken (there is no need to keep taking a drug long after it has cured you).

Normal practice with drug testing is to create two similar (randomized) groups of subjects. The *experimental group* will be given the drug being tested, while the *control group* is given a placebo. The placebo will take the form of a completely harmless and neutral 'look-alike' form of treatment (perhaps an injection of distilled water or an identical pill that contains no active drug). It is essential that the subjects are not told which group they are in, otherwise the 'placebo' effect will operate in the experimental group and not in the control group. This sort of experiment, where the subjects are not told whether they are in the experimental or the control group, is known as a *blind experiment* for obvious reasons. Of course, in certain situations, it is possible that the researcher's knowledge of which group each subject has been allocated to affects his or her behaviour in a way that influences the health of the subjects. In order to avoid this possibility, it may be a good idea to set up a *double blind experiment* where neither the experimenter nor the subjects know who is in which group.

Blind trials may be methodologically sound, but they do raise certain moral dilemmas. Consider the situation where a new drug has been discovered which appears to cure a fatal illness – perhaps terminal cancer or Aids. The volunteers who offer themselves to be tested are all desperate. This drug could be their last chance and it may seem a cruel exploitation of their plight merely to inject half of them with distilled water in the cause of adopting correct scientific procedure.

Questionnaire design

EXERCISE 8.3. A questionable questionnaire

The questionnaire below, surveying people's health practices and attitudes, is based on examples given in the book *Surveys in Social Research*, by D.A. de Vaus (George Allen & Unwin, 1986). Each question is something of a lemon in the way it is worded. Have a go at answering the questionnaire and then give some thought to the sorts of biases and inaccuracies it might produce.

A survey on health

1 How healthy are you?

2 Are the health practices in your household run on matriarchal or patriarchal lines?

3 Has it happened to you that over a long period of time, when you neither practised abstinence nor used birth control, you did not conceive?

4 How often do your parents visit the doctor?

5 Do you oppose or favour cutting health spending, even if cuts threaten the health of children and pensioners?

6 Do you agree or disagree with the following statement? 'Abortions after 28 weeks should not be decriminalized'

> **7** Do you agree or disagree with the government's policy on the funding of medical training?
>
> **8** Have you ever murdered your grandmother?
>
> ***Comments below***

As you will have discovered, this is a pretty hopeless questionnaire but there are some important general principles that can be learnt from it.

1 There is no way of knowing how to answer a question like this. There needs to be included either a clear set of words to choose from or a scale of numbers on which to rate your perception of how healthy you feel.

2 Don't ask questions that people won't understand.

3 Simplify the wording and sentence structure so that the question is clear and unambiguous without being trite or patronizing. Don't use undefined time spans (like 'a long period of time') and avoid using double negatives.

4 Avoid double-barrelled questions like this where two (or more) individuals have been collapsed into one category (in this case, 'parents'). Also, where a quantitative answer is required, give the respondent a set of categories to choose from (e.g. 'every day', 'roughly once a week', etc.). Finally, avoid gathering information from someone on behalf of someone else. They probably won't know the exact details and anyway it is none of their business.

5 This is a leading question that is pushing for a particular answer and will produce a strong degree of bias in the responses.

6 Again, avoid double negatives as they are hard to understand. This could be reworded as 'Abortions after 28 weeks should be legalized'.

7 Most people will not know what the government's policy is on the funding of medical training, so don't ask questions that people don't know about without providing further information.

8 This could be classed as a direct question on a rather sensitive issue. If you really must ask this question, there are ways of asking it more tactfully. According to de Vaus, there are four basic gambits.

(a) The 'casual' approach – 'Do you happen to have murdered your grandmother?'

(b) The 'numbered card' approach – 'Will you please read off the number of this card which corresponds with what became of your grandmother?'

(c) The 'everybody' approach – 'As you know, many people have been killing their grandmothers these days. Do you happen to have killed yours?'

(d) The 'other people' approach – 'Do you know any people who have murdered their grandmothers?' Pause for a reply and then ask, 'Do you happen to be one of them?'

There are several other aspects to designing a good questionnaire that have not emerged from this one. First and foremost, only ask questions that you want and need to know the answer to and that will provide you with information which you intend to use. Most people have only a limited capacity for filling in forms and answering questions, so don't use up their time and goodwill asking unnecessary questions. Secondly, you will need to decide whether to collect quantitative or qualitative data or a combination of the two. In making this choice, it is helpful to bear in mind how you intend, subsequently, to process the information that you collect. Qualitative information will be difficult to codify and summarize but it may allow options and opinions to be expressed that you would never otherwise hear. Quantitative data are easier to process but have the disadvantage that the range of responses is restricted to the ones you thought of when designing the questionnaire. A helpful solution to this dilemma is to offer a set of pre-defined categories but allow the respondent to add a final option of their own in the space provided, thus:

'Other, please specify' _____

Also, with questions where opinions are being sought, there may be issues on which people will have no opinion, so the option of recording 'don't know' or 'no opinion' should be offered in the range of choices.

Let us now return to an earlier suggestion that double negatives could and should be avoided by adopting a simpler wording. This is not as simple as it sounds, as the next exercise will reveal.

EXERCISE 8.4. The same but different?
Tick the boxes below to indicate whether or not you agree or disagree with these statements.

	Yes	No
a) *I do not dislike my doctor*	☐	☐
b) *I like my doctor*	☐	☐

Comments below

Two points emerge from this exercise. First, rewording a double negative may fundamentally alter the meaning of the original question. In this particular example, Statement (b) is a much stronger vote in favour of your doctor than Statement (a). Secondly, you may have observed that the options 'Yes' and 'No' are not entirely neutral words.

In conclusion, be prepared to pilot your questionnaire (try it out on a few friends first) and redraft it several times before embarking on the major survey. The layout needs to be thought about and improvements can usually be made in its general presentation. Think about the order of the questions that you have asked – is it logical and sensible? And finally, maintain respect and confidentiality for the responders and any data that you subsequently collect from them. You have both a moral and a legal duty to do so!

To end the chapter, the final exercise gives you an opportunity to put some of these ideas of good questionnaire design into practice.

EXERCISE 8.5. It's your verdict!
On 11 February 1985, under the banner headline 'We've had ENOUGH!', a tabloid newspaper produced the following questionnaire. Make a note of any criticisms of the way it was designed.

IT'S YOUR VERDICT

1 I believe that capital punishment should be brought back for the following categories of murder:

Children ☐   Police ☐   Terrorism ☐   All murders ☐

2 Life sentences for serious crimes like murder and rape should carry a minimum term of:

20 years ☐   25 years ☐

3 The prosecution should have the right of appeal against sentences they consider to be too lenient. ☐

*Tick the boxes of those statements you agree with and then post the coupon to:

VIOLENT BRITAIN, *Daily Star*, 33 St. Bride St., London EC4A 4AY.

Comments below

Summary

This chapter looked at the data collection phase of a statistical investigation. A distinction was made between *secondary* sources of data, which have already been collected by someone else, and *primary* sources of data which you collect yourself. A range of secondary sources was listed and described briefly, the most interesting and potentially useful of which you may find to be the CSO publication *Social Trends*. The section on data accuracy provided examples of two main types of error – those resulting from the inevitable limitations of the measuring tool used, and human error. The final section looked at two main issues in the area of collecting primary source data – experimental design and questionnaire design.

Comments on exercises

Exercise 8.5

The main criticism to be made of this questionnaire is the remarkable degree of restriction of the response categories offered. For example, question 2 offers only two heavy terms of sentence and there simply isn't a box for someone who would like to be 'soft on murderers and rapists' and believes that longer prison sentences won't necessarily solve the problem of violent crime. Also, the responses available in question 1 provide no opportunity for someone to express a view if they don't believe in the death penalty. Secondly, the context of the survey is highly emotive and biased, being placed within headlines such as 'We've had enough', 'My mother's killer runs free' and 'Hang the gunmen'. Perhaps it is not surprising that, from the 40 000 readers who took the trouble to complete and post the coupon, the following results were recorded.

Favouring restoring capital punishment for murder   86.33%

Favouring a 25-year minimum sentence for serious crimes of violence   92%

Favouring the right of appeal by the prosecution against sentences they consider to be too lenient   95.58%

It is interesting to contrast these findings with those of more reputable polling agencies (such as Marplan and the Prison Reform Trust, for example) who consistently report that most people, including victims of violent crime, would prefer to see compensation to victims and a system of community service rather than longer prison sentences.

9 READING TABLES OF DATA

For this chapter you will need a calculator or, if available, a computer spreadsheet.

Getting information from a table is like extracting sunlight from cucumber.

(Farquhar and Farquhar, 1891)

For many people, there is something about a table of figures that seems to make their head swim. Indeed, as one reads a page of text that contains a table of data, there is a tendency to blank the table section out; the eye almost jumps over it as if it weren't there. It is certainly true that tables can look very uninviting and incomprehensible.

The aim of this chapter is to help you to extract what information is available in a table and also to become more proficient at laying out tables of your own data so that the key features are communicated effectively. This will require a certain attitude of mind about what job it is that you think a table of figures is designed to do. It is not simply there as a way of storing information; these days we have computer data bases which will fulfil that role much more effectively. The purpose of a table, on paper, is to communicate information effectively to the person who is reading it. If it fails to do this, then it is a waste of space on the page.

Although the main thrust of this chapter deals with interpreting tables on a printed page, we need to recognize that, increasingly, this sort of data handling is done on a computer using a package known as a spreadsheet. A brief description of spreadsheets, and how to use them, is given below – if you are already familiar with spreadsheets, you may wish to skip this section.

Spreadsheets
What is a spreadsheet?

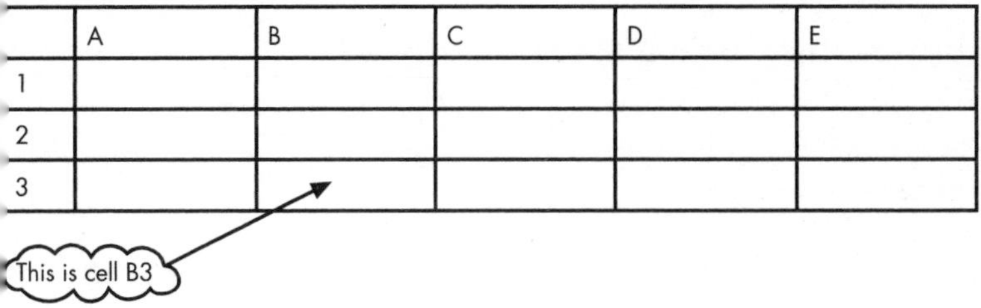

	A	B	C	D	E
1					
2					
3					

This is cell B3

Figure 9.1   Part of a spreadsheet

A spreadsheet is a computer tool which provides a way of laying out data in rows and columns on the screen. The rows are numbered 1, 2, 3, etc while the columns are labelled with letters, A, B, C, etc. Typically a spreadsheet might look something like the grid in Figure 9.1, except that rather more rows and columns are visible on the screen at any one time.

Each 'cell' is a location where information can be stored. Cells are identified by their column and row position. For example, cell B3 is indicated in Figure 9.1, and is simply the cell in the third row of column B.

The sort of information that you might wish to put into each cell will normally fall into one of three basic types – numbers, text and formulas.

■ numbers   These can be either whole numbers or decimals (e.g. 7, 120, 6.32, etc.).

■ text   These are symbols or words (for example, the headings of a row or a column of numbers will be entered as text).

■ formulas   The real power of a spreadsheet lies in its ability to handle formulas. A cell which contains a formula will display the result of a calculation based on the current values contained in other cells in the spreadsheet (for example, an average or a total or perhaps something more complicated such as a bill made up from different components). If these other component values are subsequently altered, the formula can be set up so that it automatically recalculates on the basis of the updated values and immediately gives the new result.

Why bother using a spreadsheet?

A spreadsheet is useful for storing and processing data when repeated calculations of a similar nature are required. Next to word-processing, a spreadsheet is the most frequently used computer application in business, particularly in the areas of budgeting, stock control and accounting. Spreadsheets are also being used increasingly by householders with access to a personal computer to help them to solve questions which crop up in their various roles – as shoppers, tax payers, bank account holders, members of community organizations, hobbyists, etc. It can be used to investigate questions such as:

- how much will this journey cost for different groups of people?
- is my bank statement correct?
- which of these buys is the best value for money?
- what is the calorific content of these various meals?
- what would these values look like sorted in order from smallest to biggest?
- how can I quickly express all these figures as percentages?

The reason a spreadsheet is such a powerful tool for carrying out repeated calculations is that, once it has been set up properly, you simply perform the first calculation and then a further command will complete all the other calculations automatically. Another advantage of a spreadsheet over pencil and paper is its size. The grid which appears on the screen (part of which was illustrated in Figure 9.1) is actually only a window on a much larger grid. In fact, most spreadsheets have available hundreds of rows and columns, should you need to use them. Movement around the spreadsheet is also quite straightforward – you can use certain keys to move between adjacent cells or to a particular cell location of your choice.

Using a spreadsheet

Most home computers come with a spreadsheet package bundled with the software supplied with the machine. If you have access to a computer with a spreadsheet and want to make a start at using it, then read on. In this section you will be guided through some simple spreadsheet activities.

There are many spreadsheet packages on the market and fortunately their mode of operation has become increasingly similar in recent years. However, your particular package may not work exactly as described here so you may need to be a little creative as you try the activities below.

EXERCISE 9.1. Shopping list

Load the spreadsheet package and, if there isn't already a blank sheet open, create one by clicking on File and selecting New.

Using the mouse, click on cell A1 and type in 'Milk'. Notice that the word appears on the 'formula bar' near the top of the screen. Press **Enter** (the key may be marked **Return** or have a bent arrow pointing left) and the word 'Milk' is entered into cell A1.

Using this method, enter the data below into your spreadsheet.

Milk	0.54
Bread	0.66
Eggs × 12	2.18

No comments

A very useful feature of any spreadsheet is that it can add columns (or rows) of figures. This is done by entering a formula into an appropriate cell. For most spreadsheets, formulas are created by an entry starting with '='.

EXERCISE 9.2. Finding totals using a formula

The formula for adding cell values together is '= SUM( )'. Inside the brackets are entered the cells or cell range whose values are to be added together.

Click in cell B4 and enter: = SUM(B1:B3)

Press **Enter** and the formula in cell B4 produces the sum of the values in cells B1 to B3. Now enter the word TOTAL into A4.

Your spreadsheet should now look like this.

Milk	0.54
Bread	0.66
Eggs × 12	2.18
TOTAL	3.38

No comments

Let's suppose that you gave the shopkeeper £10 to pay for these items. How much change would you expect to get? Again, this is something that the spreadsheet can do easily. Calculations involving adding, subtracting, multiplying and dividing require the use of the appropriate operation keys, respectively marked +, −, * and /. Note that the letter 'x' cannot be used for multiplication: you must use the asterisk, *. You will find these keys, along with the number keys and '=', conveniently located on the 'numeric keypad' on the right hand side of your keyboard (these keys are all also available elsewhere on the keyboard, but you may have to hunt them down, which takes time).

EXERCISE 9.3. Calculating change from £10

Enter the word 'Tendered' in cell A5, the number 10 in cell B5 and the word 'Change' in A6. Now enter a formula into B6 which calculates the change. Remember that the formula must begin with an equals sign and it must calculate the difference between the value in B6 and the value in B5.

Comments on page 160

Let's now try a slightly more complicated shopping list, this time with an extra column showing different quantities. Enter the data on page 149 into your spreadsheet, starting with the first entry in cell A8.

In order to calculate the total cost, you must first work out the cost of each item. For example, the cost of the black pens is 24 × £0.37.

Enter into cell D9 the formula: = B9 * C9

DESCRIPTION	QUANTITY	LIST PRICE	COST
Pens black	24	0.37	
Folders	45	0.25	
Plastic tape	4	0.68	
		TOTAL	

This gives a cost, for the pens, of £8.88. You could repeat the same procedure separately for the folders and the plastic tape but there is an easier way, using a powerful spreadsheet feature called 'fill down'. You 'fill down' the formula currently in D9 so that it is copied into D10 and D11. The spreadsheet will automatically update the references for these new cells.

Click on cell D9 (which currently displays 8.88) and release the mouse button. Now move the cursor close to the bottom right hand corner of cell D9 and you will see the cursor change shape (it may change to a small black cross, for example). With the cursor displaying this new shape, click and drag the mouse to highlight cells D9–D11 and then release the mouse button. The correct costs for the folders (£11.25) and the plastic tape (£2.72) should now be displayed in cells D10 and D11 respectively. Now click on cell D10 and check on the formula bar at the top of the screen that the cell references are correct. Repeat the same procedure for cell D11. Magic!

EXERCISE 9.4. Summing up

With the costs in column D completed, you are now able to calculate the total costs. As before, this requires an appropriate entry in D12 using the = SUM command. Do this now; you should get a total bill of £22.85.

No comments

The data in Table 9.1 (page 150) show the number (in thousands) of notifiable offences recorded by the police, by type of offence, in England and Wales in 1981 and 1996.

NOTIFIABLE OFFENCES	1981	1996
Theft of vehicles	333	493
Theft from vehicles	380	800
Burglary	718	1165
Violence against the person	100	239
Fraud and forgery	107	136
Sexual offences	19	31

Table 9.1   Notifiable offences, 1981 and 1996
Source: *Social Trends 28*, Table 9.4

EXERCISE 9.5. Calculating percentages

Enter the data in Table 9.1 into your spreadsheet, starting with the first entry in cell A14. By the way, you will find that column A is not wide enough to accommodate all the text. Don't worry about this for the moment. When all the data have been entered, move the cursor to the top of the spreadsheet just above cells A1 and B1. As the cursor moves to a position close to the vertical line between the column headings A and B, it changes shape. Click on this vertical line and drag to the right to widen column A. Adjust the width of this column until it accommodates all the text.

Using 'fill down', calculate the percentage increase in each category of crime over the period. Which crime had the greatest and which had the least percentage increase?

Comments on page 161

What else will a spreadsheet do?

This brief introduction has only really scratched the surface of what can be done with a spreadsheet. For a start, three tiny data sets shown here were merely illustrative and don't properly reveal the power of a spreadsheet. Suppose, on the other hand, you have a table with 200 rows of data. Instead of having to perform 200 separate calculations you only have to do one and you can 'fill down' the rest.

Once data have been entered into a spreadsheet, there are many options available for seeking out some of the underlying patterns. For example, columns or rows can be re-ordered or sorted either alphabetically or according to size. As you have already seen, column and row totals can be inserted. Also, you can easily divide one column of figures by numbers in an adjacent column to calculate relative amounts or to convert a set of numbers to percentages.

As well as the four operations $+$, $-$, $\times$ and $\div$ which can be found directly on the keyboard, a variety of other functions will also be contained within one of the spreadsheet's many menu options. These will enable you to select some or all of the data and find the mean, median, maximum value, standard deviation, and much more, at the touch of a button. For example, try clicking in cell D13 and enter the formula = AVERAGE(D9:D11). This should give you the mean of the numbers in the three cells D9, D10 and D11 (not a particularly helpful figure, as it happens, but you get the general idea!). The good news is that you won't have to remember the various commands needed to calculate these sorts of summary; they can be pasted directly from the appropriate menu option (possibly named 'Function' or something similar). An advantage of selecting functions via the menu is that they will be made available in a user-friendly way so that the command syntax is made apparent.

Finally, most spreadsheets have powerful graphing facilities which also allow you to select either all or some of the data and display them as a pie chart, bar chart, scattergraph, line graph, and so on. The detailed operation of the graphing facilities varies considerably from one spreadsheet package to another so they are not explained here. However, if by now you have greater confidence with using a spreadsheet, this is something that you might like to try for yourself. When you feel ready, return to Chapters 3 and 4 and see if you can reproduce the graphs displayed there. You could also return to Chapter 5, *Summarizing data* and try your hand at calculating means, medians, standard deviations and so on using the spreadsheet.

Turning the tables

In this section you will look at the way in which information is often set out in a table and how, with a little common sense, it might be done better. We start with a set of data in Table 9.2 which shows the population and urban population in European Union countries and other selected regions.

COUNTRY/ REGION	POPULATION (M) 1997	POPULATION (M) 2025	URBAN POPULATION (M) 1995
Africa	758.4	1453.9	257.9
Australia	18.3	23.9	15.6
Austria	8.2	8.3	4.6
Belgium	10.2	10.3	9.9
Canada	29.9	36.4	23.0
China	1244	1480	373.0
Denmark	5.2	5.3	4.4
Europe	729	701	540
Finland	5.1	5.3	3.2
France	58.5	60.4	42.7
Germany	82.2	80.9	71.5
Greece	10.5	10.1	6.8
Ireland	3.6	3.7	2.1
Italy	57.2	51.7	38.3
Luxembourg	0.4	0.5	0.4
Netherlands	15.7	16.1	14.0
Portugal	9.8	9.4	3.5
Spain	39.7	37.5	30.2
Sweden	8.8	9.5	7.3
United Kingdom	58.2	59.5	51.8
United States	271.6	332.5	206.4

Table 9.2   Population and urban population of European Union countries and other selected regions
Source: Social Trends 28, 1998, adapted from Table 1.20

EXERCISE 9.6. What is it telling me?

Look at Table 9.2 and think about the following questions.

a) Do you feel that the layout of the table could be improved?

b) What sort of information does it reveal about urban population in European Union countries and other selected regions?

c) What additional information might you need or what further calculations might you wish to do in order to squeeze more useful information from it?

Comments follow

A confusion here is that the countries and regions have been presented in alphabetical order, with the result that the European Union countries are mixed up with countries and regions from elsewhere. Another criticism of the layout is the rather untidy way that the figures have been written down – the data for Europe have been presented as whole numbers while others are given to one decimal place of accuracy. This causes two problems. Firstly it creates an inconsistency between the degree of accuracy to which the data from each country has been stated – the statement that 'the population of Europe in 1997 was 729 million' is rather different from saying it was, say 729.0 million. The second problem is that this inconsistency has the effect of throwing the numbers out of alignment, thereby making it awkward to run your eye down a column of figures and compare countries. But we shall tackle these problems of layout later. Let us start by trying to make sure that the information being presented is the most useful that is available.

You may have felt that, as it stands, Table 9.2 doesn't tell you much. Regions such as China, Africa and Europe have the largest urban populations, but this is hardly surprising since these regions have also the largest overall populations. Also, the information, even at the time of writing, is out of date – in 1997 there were 15 countries in the European Union. How many are there now? How many will there be in one, two, five years' time? Having said that, there is always a time lag between the collection and the organization of data and an even longer delay until their publication, so *it is always the case that data are out of date!*

A more sensible way of comparing urban populations across these very different regions is to concentrate not on the *absolute size* of the urban population but instead to find the *percentage* of the population who live in urban areas. The first problem is that the years don't quite match up. However, the years corresponding to columns 2 and 4 of the table (i.e. for 1997 and 1995, respectively) are close enough for our purposes. Let us take as an example the data for the first region in the table, Africa. To calculate the percentage of the 1997 population that lived in urban areas, calculate:

$$\frac{257.9}{758.4} \times 100$$

This gives an answer of 34.005 801 69.

Since the original data were given to one decimal place of accuracy, it makes sense to round this result to one decimal place, giving 34.0%.

COUNTRY/ REGION	POPULATION (m) 1997	POPULATION (m) 2025	URBAN POPULATION 1995	URBAN POPULATION (%)
Africa	758.4	1453.9	257.9	34.0
Australia	18.3	23.9	15.6	85.2
Austria	8.2	8.3	4.6	56.1
Belgium	10.2	10.3	9.9	97.1
Canada	29.9	36.4	23.0	76.9
China	1244	1480	373.0	30.0
Denmark	5.2	5.3	4.4	84.6
Europe	729	701	540	74.0
Finland	5.1	5.3	3.2	62.7
France	58.5	60.4	42.7	73.0
Germany	82.2	80.9	71.5	87.0
Greece	10.5	10.1	6.8	64.8
Ireland	3.6	3.7	2.1	58.3
Italy	57.2	51.7	38.3	67.0
Luxembourg	0.4	0.5	0.4	100.0
Netherlands	15.7	16.1	14.0	89.2
Portugal	9.8	9.4	3.5	35.7
Spain	39.7	37.5	30.2	76.1
Sweden	8.8	9.5	7.3	83.0
United Kingdom	58.2	59.5	51.8	89.0
United States	271.6	332.5	206.4	76.0

Table 9.3   Percentage urban population

We now need to perform the same calculation for each country or region. Phew! Although it sounds a daunting prospect, rest assured that I have already entered the data onto a spreadsheet and this sort of calculation is just what spreadsheets do best – in this case, dividing a column of figures by the corresponding figures in another column, and multiplying each result by 100. I have also ensured that each result has been corrected to the same number of decimal places – I returned to the original source and amended the data for Europe to one decimal place. As was mentioned earlier, not only does this ensure that the data are internally consistent but also the columns of figures are now properly aligned, making for easier comparisons as you run your eye down a particular column. Table 9.3 shows the results of this operation.

EXERCISE 9.7. Spreadsheet (or calculator) check

If you have access to a spreadsheet, copy the data from the first four columns of Table 9.3 into columns A, B and C. Perform an appropriate calculation in Column D and 'fill down' to calculate the percentage urban data. Then check your results against the final column in Table 9.3 to see that you have done the calculation correctly.

If you don't have access to a spreadsheet, you can use your calculator to confirm your understanding of this calculation by checking some of the values in the final column of the table.

Comments on page 161

Now that we have the data we want, the next stage is to try to improve the presentation and layout. Table 9.4 has been 'tweaked' by having the EU countries separated from the rest.

And now it is time for the final reorganization. One or two further 'tweaks' are worth thinking about.

Firstly, notice that the EU countries have been presented in alphabetical

COUNTRY/ REGION	POPULATION (m) 1997	POPULATION (m) 2025	URBAN POPULATION 1995	URBAN POPULATION 1995 (%)
Austria	8.2	8.3	4.6	56.1
Belgium	10.2	10.3	9.9	97.1
Denmark	5.2	5.3	4.4	84.6
Finland	5.1	5.3	3.2	62.7
France	58.5	60.4	42.7	73.0
Germany	82.2	80.9	71.5	87.0
Greece	10.5	10.1	6.8	64.8
Ireland	3.6	3.7	2.1	58.3
Italy	57.2	51.7	38.3	67.0
Luxembourg	0.4	0.5	0.4	100.0
Netherlands	15.7	16.1	14.0	89.2
Portugal	9.8	9.4	3.5	35.7
Spain	39.7	37.5	30.2	76.1
Sweden	8.8	9.5	7.3	83.0
United Kingdom	58.2	59.5	51.8	89.0
Europe	729.2	701.1	539.6	74.0
Canada	29.9	36.4	23.0	76.9
United States	271.6	332.5	206.4	76.0
Africa	758.4	1453.9	257.9	34.0
China	1243.7	1480.4	373.1	30.0
Australia	18.3	23.9	15.6	85.2

Table 9.4   Populations of EU countries separated from other selected countries/regions

order. While this is fine if you want to find the data for a particular country quickly, it is less than fine for seeing overall patterns in the data. A more useful strategy is to order the countries according to the size of the measure that you are most interested in. Unless there are other pressing reasons determining the order in which the rows or columns are to be presented, it is sensible to put the most important or interesting ones first. Let us assume, therefore, that it is the percentage urban population that is the particular factor of interest. I propose to sort this column so that it starts with the 'biggest first' on the grounds that the biggest values are usually the most interesting and the top part of the table is probably where most people look most carefully. Clearly, all the other columns have to follow suit so that the data for each country match up. On a spreadsheet, this is achieved using the **Sort** command. When using **Sort** on a table like this, containing several linked columns, you need to be careful not to scramble the data. For example, if you were to sort one of these columns on its own, the data would no longer match up with the data in the other columns. Before sorting, you must therefore select the entire table and then carry out the sort by column B. This ensures that the rest of the table is also sorted in the same way. As you will see from looking at Table 9.5, I sorted the EU countries and then the other regions separately.

Secondly, it is now time to reassess whether the one decimal place accuracy is the most appropriate for our purposes. Remember that we originally settled on one decimal place accuracy because it could be justified in terms of the accuracy of the data from which the rates were calculated. But it might well be that this level of accuracy is actually too great for the main purpose for which this table has been constructed, namely to look for interesting patterns. At this stage you could round the data to the nearest whole number. Table 9.5, which will be my final version, shows the data re-ordered for ease of interpretation.

Finally, with any set of data it is always worth bearing in mind what a typical or middle value looks like and how the others compare to it. Thus an extra row showing the median values for the EU countries has been added directly below the final EU entry. An italic type-face has been used to distinguish it from the other countries and the gap below it has been kept so that there is a clear separation between the two sets of countries/regions.

It is worth pointing out that I have presented this example in four separ-

COUNTRY/ REGION	POPULATION (m) 1997	POPULATION (m) 2025	URBAN POPULATION 1995	URBAN POPULATION 1995 (%)
Luxembourg	0.4	0.5	0.4	100
Belgium	10.2	10.3	9.9	97
Netherlands	15.7	16.1	14.0	89
United Kingdom	58.2	59.5	51.8	89
Germany	82.2	80.9	71.5	87
Denmark	5.2	5.3	4.4	85
Sweden	8.8	9.5	7.3	83
Spain	39.7	37.5	30.2	76
France	58.5	60.4	42.7	73
Italy	57.2	51.7	38.3	67
Greece	10.5	10.1	6.8	65
Finland	5.1	5.3	3.2	63
Ireland	3.6	3.7	2.1	58
Austria	8.2	8.3	4.6	56
Portugal	9.8	9.4	3.5	36
Median EU	*10.2*	*10.1*	*7.3*	*76*
Australia	18.3	23.9	15.6	85
Canada	29.9	36.4	23.0	77
United States	271.6	332.5	206.4	76
Europe	729.2	701.1	539.6	74
Africa	758.4	1453.9	257.9	34
China	1243.7	1480.4	373.1	30

Table 9.5   Population and urban population of European Union countries and other

ate and rather drawn-out stages, In practice, or course, several of these stages would be collapsed into one. A great virtue of performing the task on a spreadsheet is that you can try out many possible layouts very easily before settling on the one which is the most effective.

Now that the table has been reorganized to our satisfaction, let us return to the central question on which this exercise in data interpretation was based.

> What sort of information does it reveal about urban population in European Union countries and other selected regions?

What are the main conclusions to be drawn from running an eye down the final column of Table 9.5? Perhaps one overall theme is that, in the developed world (Europe, USA, Canada and Australia) levels of urbanization are consistently high. So it would seem that the wide-open spaces of Canada and Australia are clearly home to a relatively small number of their citizens.

The key contrast is with Africa and China, which still have largely rural economies. You may wonder what other economic indicators are linked to high levels of urbanization; it could be the case that, in general, wealthier countries also tend to be more urbanized, but it would require more data and further investigation to confirm this.

Your analysis has also revealed that the most highly urbanized parts of the EU are the Benelux countries – Luxembourg, Belgium and the Netherlands. Very much on its own is Portugal with a markedly different pattern of urbanization from the rest of the EU. Again this raises further questions about why, which are not answerable from the data provided here.

To end this section, here are two additional handy hints concerning the layout and presentation of a table of data.

- For certain types of tables it may be useful to add together all the row totals and also add together the column totals. The sum of the row totals should equal the sum of the column totals and this can provide a check for internal consistency.

- Trying to read a table which contains too many rows or columns can be overwhelming, in which case it may be possible to collapse some of the rows or columns together to make a smaller, simpler table.

Reading the small print

If people's eyes tend to blank out tables of figures, you can be darn sure that they blank out the small writing that goes around them.

(Alan Graham, 1999)

Reading tables of data intelligently can be quite a creative exercise, drawing on the use of a range of detection skills in order to gain some better understanding of the story that lies behind the figures. So far the focus has been entirely on the main body of the table and how we might reorganize or recalculate the data that it contains. However, there are often many other vital – but less visible – clues around the edges. In particular, the title, source and footnotes may provide essential background information which may require you to qualify or refine any conclusions you make.

GREAT BRITAIN	PERCENTAGES				
	1961	1971	1981	1991	1996–7
One person	14	18	22	27	27
Two people	30	32	32	34	34
Three people	23	19	17	16	16
Four people	18	17	18	16	15
Five people	9	8	7	5	5
Six or more people	7	6	4	2	2
All households (= 100%) (millions)	16.3	18.6	20.2	22.4	23.5
Average household size (number of people)	3.1	2.9	2.7	2.5	2.4

Table 9.6   Households by size
Source: Adapted from *Social Trends 28*, 1998, Table 2.2

Finally a few words about title and source. A good title should be both clear and concise. Sometimes it is hard to be both of these things at the same time, in which case, be concise and put any additional qualifications or details in a footnote. A common area of confusion in titles dealing with British data is whether the data refer to the UK, to Great Britain or to England & Wales. These are all different land masses with different (though overlapping) populations and careful reading of the title should reveal which you are dealing with. The importance of giving a source is to enable the reader to check your data and perhaps to chase up further information about them.

EXERCISE 9.8. What does it all mean?
Look at Table 9.6.

  a) Which parts of the United Kingdom are included and which are excluded in these data?
  b) Why might the term 'household size' be ambiguous?
  c) Using a spreadsheet, confirm that the data on 'Average household size' in the final row of the table is consistent with the rest of the table.

Comments on page 162

Summary

This chapter looked at different aspects of the design of tables of data. The first section explained how this can be done on a computer using an office software package called a spreadsheet. Some useful strategies which were mentioned for tidying up and simplifying tables included the following.

■ Remove unnecessary rows and columns (too much information is off-putting).

■ Remove unnecessary digits (too much accuracy is off-putting).

■ Re-order rows and columns where necessary in order to help the main features stand out.

■ Append row and column totals and/or averages. The totals can be used as a check for internal consistency as well as giving an overview.

■ Raw 'absolute' data may be unsuitable for making comparisons, in which case it may be possible to change them into a 'relative' form – such as rates or percentages.

■ Round the data to as few figures as you dare. Remember that the central purpose of the table is less to store data than to reveal patterns that they might contain.

■ The table may not provide all the information you need, in which case you may need to check the original source or chase up a further source.

■ Read the small print carefully, especially the title, source and footnotes.

Comments on exercises

Exercise 9.3
The required formula, which completes the table below is: $= B5 - B4$
Your final spreadsheet table should look like this.

Milk	0.54
Bread	0.66
Eggs × 12	2.18
TOTAL	3.38
TENDERED	10
CHANGE	6.62

Exercise 9.5

In order to calculate the percentage increase of thefts of vehicles between 1981 and 1996, you must first find, in thousands the actual increase, which is $493 - 333 = 160$. To convert this to a percentage, this figure is divided by 333 and the answer multiplied by 100. Expressed as a single calculation, this is:

$$\frac{493 - 333}{333} \times 100$$

On a spreadsheet, the corresponding formula (keyed into cell D15) is:

$$= (C15 - B15)/B15*100$$

After 'filling down', the completed table should look like this.

NOTIFIABLE OFFENCES	1981	1996	% INCREASE
Theft of vehicles	333	493	48.0
Theft from vehicles	380	800	110.5
Burglary	718	1165	62.3
Violence against the person	100	239	139.0
Fraud and forgery	107	136	27.1
Sexual offences	19	31	63.2

Largest percentage increase: violence against the person (139.0%)

Smallest percentage increase: fraud and forgery (27.1%)

Exercise 9.7

The spreadsheet formula for cell E2 was:

$$= D2/B2*100$$

Exercise 9.8

    (a) Great Britain includes England, Scotland and Wales. The United Kingdom covers these three countries plus Northern Ireland, so Northern Ireland is excluded from the data in Table 9.6.

    (b) It isn't obvious exactly what types of social groupings this term 'household' covers. For example, would it include large communal houses such as boarding schools or homes for the aged? It is possible that some countries define 'households' differently to others, in which case international comparisons become rather dubious. For the purpose of this survey it is defined in terms of people living together in a domestic setting and sharing facilities.

    (c) A spreadsheet approach to checking the final row of data in the table is suggested below.

As an example, here is one way of checking the figure for 'Average household size of 2.4' for 1996–97. Firstly, Column A needs to be replaced by numbers, 1, 2, 3, etc. A problem occurs with the final category of household size, 'Six or more people', which could include 6, 7, 8, ... people up to anything like 15 or 20. How can we take a representative number for this category? This problem was discussed in Chapter 5, *Summarizing data* in the calculation of a weighted mean. This time, since the number of very large households is much lower than before, we shall adopt a smaller household size of 7 to represent this 'Six or more people' category.

Enter the data on page 163 into Columns A to C of a blank spreadsheet.

	SIZE	1996–97
One person	1	27
Two people	2	34
Three people	3	16
Four people	4	15
Five people	5	5
Six or more people	7	2

You now use the spreadsheet to calculate the weighted mean of the data in Columns B and C. In cell D2 enter the following formula:

$$= B2 * C2/100$$

Then fill down from D2 to D7. Click in D8 and enter the following formula:

$$= SUM (D2:D7)$$

This produces the weighted mean of the values in column B, weighted by the values in column C, of 2.42. Your final spreadsheet table should look like this.

	SIZE	1996–97	
One person	1	27	0.27
Two people	2	34	0.68
Three people	3	16	0.48
Four people	4	15	0.6
Five people	5	5	0.25
Six or more people	7	2	0.14
		Weighted mean 2.42	

The answer of 2.42 is consistent with the value of 2.4 shown in Table 9.6 (rounded to one decimal place).

10 REGRESSION: DESCRIBING RELATIONSHIPS BETWEEN THINGS

You will need a calculator for this chapter.

Paired data

Data collection generally takes place from a sample of people or items. Depending on the purpose of the exercise, we often take just *one single* measure from each person or item in the sample. For example, a nurse might wish to take a group of patients' temperatures, or a health inspector might measure the level of pollution at various locations along a river. These sorts of sample will consist of a set of single measurements which should help to provide a picture of the variable in question. So, the nurse could use the temperature data to gain some insight into the average temperatures of patients in hospital and how widely they are spread. Similarly, the health inspector could use the pollution data to establish the typical levels of pollution present in the river, and also see to what extent the levels of pollution vary from place to place.

Sometimes, however, data can be collected *in pairs*. Paired data allows us to explore something quite different than is indicated by the examples above. Using paired data, we can look at the possible *relationship* between the two things being measured. To take a simple example, look at the two graphs, Figure 10.1, which show the typical amounts of sunshine and rainfall over the year in the Lake District.

EXERCISE 10.1. Hot dry summers?
  a) *How would you describe the pattern of sunshine over the year?*

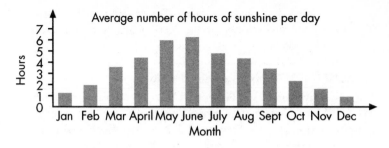

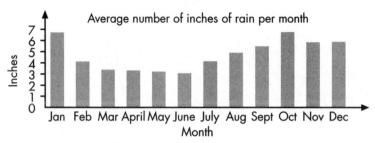

Figure 10.1   Weather in the Lake District
Various sources

---

b) *How would you describe the pattern of rainfall over the year?*

c) *Putting a) and b) together, is there any pattern in the relationship between the amount of sunshine and the amount of rainfall over the year?*

Comments below

(a)   Generally, the sunniest months are in the middle of the year, around May and June, while the months at the beginning and end of the year have the least sunshine.

(b)   Rainfall is least in the middle of the year, around May and June, while the months at the beginning and end of the year have most rainfall.

(c)   It seems clear from (a) and (b) above that there *is* a relationship between rainfall and sunshine. In general, the months which have the most sunshine tend to have the least rainfall, and vice versa. The pattern in this relationship can be spelt out more clearly when drawn on a scattergraph.

EXERCISE 10.2. Graphing paired data

Use the graphs in Figure 10.1 to complete Table 10.1. Then plot these pairs of data on the scattergraph provided in Figure 10.2 – the first point has already been plotted for you. (If you feel it necessary, reread the section on graphs in Chapter 2.)

MONTH	AVERAGE HOURS OF SUNSHINE (ESTIMATED)	RAIN (ESTIMATED)
January	1.0	6.6
February		
March		
April		
May		
June		
July		
August		
September		
October		
November		
December		

Table 10.1   Data from the graphs in Figure 10.1

Comments on page 181

If you are in any doubt about how to plot the points, you can check your scattergraph with the one given on page 182. As you should see from your scattergraph, the pattern of points is fairly clear. The points seem to lie roughly on a straight line which slopes downwards to the right. This bears out the conclusions given to Exercise 10.1 part (c), namely that high sunshine tends to be associated with low rainfall. The main ideas of this chapter and the one which follows it are all about *interpreting more precisely the sort of patterns that can be found on a scattergraph.* A key point to remember, however, is that these ideas involve investigating relationships between two things and therefore throughout this chapter we will be dealing with *paired data.*

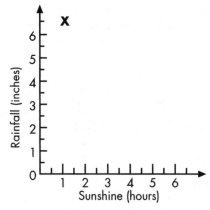

Figure 10.2 Scattergraph showing the relationship between the amount of sunshine and rainfall over a typical year
Source: Data from Table 10.1

The 'best-fit' line

The word used to describe the overall shape of points plotted on a scattergraph is the *trend*. It seems clear from the previous example that the overall trend in Figure 10.2 is a downward-sloping one. However, it is possible to be a bit more precise about this pattern. Imagine drawing a line through the middle of these points. Finding the equation of such a line would allow us to move from vague, wordy descriptions of the trend to a more precise mathematical description of the relationship in question. Once the line has been defined in this more formal way, it becomes possible to make predictions about where we think other points might lie. Incidentally, this involves using a few basic mathematical ideas about graphs and co-ordinates and these are explained in Chapter 2.

The 'best-fit' line or, *regression line*, as it is sometimes called, is the best line which can be made to fit the points which it passes through. However, a straight line may be rather too crude a description of the trend. In the world of economic and industrial forecasting, for example, curves are commonly used and these provide a necessary degree of subtlety and accuracy. For instance, economic data are often subject to seasonal fluctuations (people tend to spend more money on household goods around December and January and less in the summer). Fitting straight lines to these sorts of data would be to ignore this seasonal factor and result in highly inaccurate predictions. But for the purposes of this chapter, we will look only at methods using straight lines. This is called *linear* (i.e. straight line) regression.

EXERCISE 10.3. By eye
   a) *Using a ruler and pencil, draw by eye the 'best-fit' straight line onto the scattergraph in Figure 10.2.*
   b) *Think about how you might describe this line mathematically.*

Comments below

In order to define a straight line exactly, two pieces of information are required. By convention, these are usually the *slope* of the line, and the point on the graph where it crosses the vertical axis (known as the *intercept*). The slope can be either a positive or a negative number. Lines which slope from bottom left to top right will have a positive slope – i.e. the value of Y will increase as the value of X increases. Lines which have a negative slope run from top left to bottom right. These ideas are explained in Chapter 2 and if you are unfamiliar with them or find them confusing, you should go back and reread the section 'Equations and graphs' now.

In this example, Rainfall is on the vertical axis and Sunshine on the horizontal axis. The conventional labels for these axes are, respectively, Y and X. In general, if we are seeking to find the linear regression line, the equation of the straight line required will be of the form:

$$Y = a + bX$$

where Y represents Rainfall (measured in inches per month)
        X represents Sunshine (measured in hours per day)

and a and b are the two pieces of information required to define the line,
i.e. a = intercept
       b = the slope.

Incidentally, as was explained in Chapter 2, there are several possible ways of setting up this equation. Firstly, you might prefer to use labels

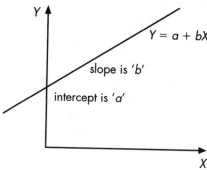

Figure 10.3 The regression coefficients

which are easier to remember than X and Y; R and S, for instance. If so, the regression equation would look like this:

$$R = a + bS^1$$

a and b are known as the *regression coefficients*.

Now you have drawn in the best-fit line by eye (Exercise 10.3, part a) the next stage is to find its equation. The intercept, which is the 'a' value in the equation, is quite easy to read off the graph, but the slope (the value of 'b' in the equation) is less straightforward. Try it yourself first in Exercise 10.4 (page 170).

In practice, best-fit equations are rarely found in this way but Exercise 10.4 was included to give you an intuitive feeling for what is involved in calculating regression coefficients. There are several other methods for finding the values of a and b, the simplest of which is to feed the raw data directly into a suitable calculator (or microcomputer) and obtain them at the press of a button. When all the data have been entered, you then simply press the keys marked 'a' and 'b' to discover the values for the intercept and slope of the best-fit line. If you do have a suitable calculator which handles linear regression, spend a few minutes now entering the paired data given in Table 10.1. If you have never done this before, be prepared to spend some time reading through your calculator manual.

---

[1] You may be more familiar with straight line equations which have the form '$Y = mX + c$', than the one given here. However, in statistics, the form '$Y = a + bX$' is the more common and, if you have a calculator which finds regression lines, the letters 'a' and 'b' are likely to be the labels used on the relevant keys.

EXERCISE 10.4. Finding the regression coefficients, *a* and *b*, from the graph

a) *Read off the intercept where the line passes through the vertical axis. This is the 'a' value in the equation of the best-fit line.*

b) *Remember that the slope of a line is the amount it increases for each unit it travels horizontally from left to right (a decrease is measured as a negative quantity). From the graph estimate the slope, i.e. the 'b' value in the best-fit line.*

c) *Using these estimates for 'a' and 'b', write down the equation of the 'best-fit' line.*

Comments on page 182

Alternatively, if you do not have a calculator which will calculate the regression coefficients, there is a formula which allows you to calculate them directly. This is included on pages 179–181, so turn to this section now if you don't have a suitable calculator or wish to see how to calculate the regression coefficients using the formula. If you aren't interested in the mathematical aspects of calculating the values of *a* and *b*, you might prefer to omit this section. However, it would still be helpful to have a look at Figure 10.9 on page 182, which shows the best-fit line and how the values of *a* and *b* show up on the graph.

In this example the slope had a negative value (-0.596), which was reflected in a downward-sloping line. This type of trend suggests a *negative relationship* between sunshine and rainfall (i.e. as one increases, the other tends to decrease). Where the slope is positive, we say that there is a positive relationship between the two things under consideration (i.e. as one increases, the other also tends to increase).

At the level at which this book is aimed, it is enough to know what the regression coefficients mean (i.e. that they give the slope and the intercept of the regression line) and to be able to calculate them, preferably using a suitable calculator. However, it is also worth knowing that the 'best-fit line' through a set of points has one important property, namely that if you take the distances of each point from this line, square them and add the squares together, the result is smaller than for any other straight line you might have chosen. In fact, the formula given in the Comments derives from this so-called 'least squares' property. For this reason, then, this line of best-fit is also known as the 'least-squares' line. In other words, the following terms all mean more or less the same thing:

- Best-fit line
- Least squares line
- Linear regression line

The next exercise will give you the chance to calculate a best-fit line for yourself and also to give some thought as to how it might be interpreted.

EXERCISE 10.5. Family planning

Table 10.2 gives the number of family planning clinic sessions over an eight year period, along with the conception rate for women under 20 for the same period.

YEAR	FAMILY PLANNING CLINIC SESSIONS (000) (X)	CONCEPTION RATE (PER 1000) (Y)
1	202	58
2	197	57
3	197	56
4	199	56
5	196	58
6	195	61
7	194	62
8	191	66

Table 10.2   Family planning
Source: Estimated from UK national data

a) *Plot the data on the scattergraph, Figure 10.4. Estimate by eye the slope and intercept from the graph. (In this example it is quite difficult to estimate the intercept by eye, due to the fact that the X axis is not zeroed at the origin.)*
b) *Using a spreadsheet, a statistical calculator, or by the long method given on pages 179–181, find the equation of the best-fit line. Draw it on your graph and check that it actually looks like a 'line of best fit'.*

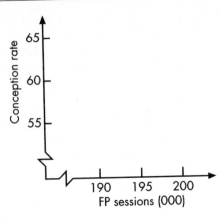

Figure 10.4

    c) *Does the best-fit line have a positive or a negative slope? What does this mean in terms of these two variables?*

Comments on page 183

Making predictions

By now you should have mastered the technicalities of calculating the regression coefficients and drawing the best-fit line through a scatter of points. Essentially, this best-fit line is a mathematical summary of what you think might be the relationship between the two variables in question. However, what is more interesting than calculating it is to use and interpret it. What we can now do is to use the line to make predictions about other pairs of values, perhaps by extending the line or by making guesses about points on the line where data are absent. However, it is worth stressing that these really are only guesses which carry all sorts of assumptions that may not hold in practice. Now do Exercise 10.6.

As you will have read in the comments to this exercise, there are considerable dangers in using a regression line to make estimates of other values. In general, these dangers are least when estimates are made within the range of values covered by the sample – for example in the case of the prediction made about Mia's pulse rate. Predictions of this sort, which lie within the range of the sample, are called *interpolation*

EXERCISE 10.6. Racing pulses?

The pulse rates and ages have been recorded for fourteen people and are shown in Table 10.3.

a) *Plot the data onto the scattergraph provided in Figure 10.5. What does the pattern of points suggest to you about the relationship between someone's age and their pulse rate?*

b) *Either by eye, or by calculating the regression coefficients, draw the linear regression line (i.e. the best-fit line) through your points.*

c) *Use the regression line to make the following predictions.*

- *Mia is 60 years old. What pulse rate would you expect her to have?*
- *Mayerling is two years old. What pulse rate would you expect her to have?*
- *Tom has a pulse rate of 90 beats per minute. How old do you think he is?*

d) *What assumptions have you made in carrying out these predictions?*

Comments on page 185

(literally 'within the points'). Of course, even interpolated estimates may be extremely inaccurate for a number of reasons – the regression line is only an average, after all, and merely gives rough predictions. Thus, it would be silly to quote Mia's predicted pulse rate, say, to two decimal places, as 72.04. A prediction of 'about 72' or even 'about 70' would be more realistic.

The dangers of inaccuracy increase when making predictions which involve having to extend the line *beyond* the range of values in the sample. Such predictions are known as *extrapolation* (literally going 'beyond the points'), and the estimates made about Tom and Mayerling are examples of extrapolating or extending the regression line. In this example, both these predictions are almost certainly wildly inaccurate. In the case of Tom, most people aged 138 tend to have a pulse rate much closer to zero beats per minute than 90! What is more likely is that Tom

NAME	AGE (YRS)	PULSE RATE (BEATS PER MINUTE)
Ellen	33	62
Paul	50	66
Ramesh	45	77
Karl	65	78
Donna	57	78
Eliza	70	70
Siobhan	42	62
Joseph	75	76
Joanne	57	67
Debbie	35	70
Indira	44	68
Sean	63	70
Beena	50	70
Tara	55	72

Table 10.3 The ages and pulse rates of a sample of 14 adults
Source: Personal survey

had just completed a bout of intensive exercise prior to having had his pulse taken. Predicting Mayerling's pulse rate involved extrapolating in the other direction. As before, we have assumed that the pattern which seems to be true for the sample of 14 older people also holds true for children. Unfortunately, it is just not the case. In fact, for children, the relationship between age and pulse rate tends to be a negative, rather than a positive one, as the data in Table 10.4 and Figure 10.6 illustrate.

Seeing these additional data may now encourage you to reconsider the case of Tom. On the basis of this second regression line, someone with a

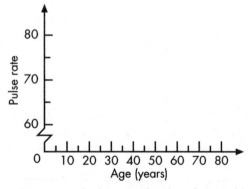

Figure 10.5   Scattergraph showing the data from Table 10.3

NAME	AGE (YRS)	PULSE RATE (BEATS PER MINUTE)
Irena	10	86
Nadia	3	107
Luke	4	98
Jon	6	105
Carrie	8	90
Theo	1	115
Jagjeet	16	69
Ira	21	81
Hannah	5	100
Alex	2	105

Table 10.4 The ages and pulse rates of a sample of 10 young people
Source: Personal survey

regular pulse rate of 90 beats per minute might be expected to be 10 or 11 years of age.

Taken separately the pulse rate data from these two separate samples seemed to indicate that linear regression was a valid way of summarizing the trends in each case. However, when we start to consider the two samples taken together – that is to look at the relationship between pulse rate and age over the entire age range – we begin to question whether linear regression is a sensible way of describing what we see. If the two samples are collapsed together and the entire 24 points are plotted onto the same graph, a very different picture emerges, see Figure 10.7.

There are methods for finding a non-linear curve which fits this pattern – a sort of 'curve of best-fit' – but these are beyond the scope of this book.

Now to end this section on regression lines, Exercise 10.7 allows you to think more deeply about aspects of regression which may have concerned you.

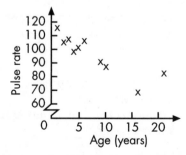

Figure 10.6 Scattergraph showing the data from Table 10.4

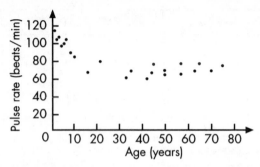

Figure 10.7   Scattergraph showing the data from Tables 10.3 and 10.4 together

---

EXERCISE 10.7. Some further questions about the line of best fit

a) A dictionary definition of the word 'regression' will be something like 'going back to an earlier or more primitive form'. How does this tie in with what you have read in this chapter about regression in statistics?

b) Suppose you wish to calculate a regression line where one of the chosen variables was 'year'. Do you think it would be necessary to enter each year in full (e.g. 1989, 1990, 1991, . . .) or could you enter them as 89, 90, 91, etc?

c) If the scales on either of the two axes are changed, will this alter the slope of the best-fit line?

d) So far, the labels for the axes of each of the scattergraphs have been chosen for you, so you have not had to choose which way round they should go. Do you think it matters which way round they are?

e) What are the most common types of data for which regression lines are used?

f) How good do you think the fit has to be for the best-fit line to be meaningful?

Comments follow

(a) There is a clear link between the statistical meaning of 'regression' and its everyday use. The regression line can be thought of as a simplified (or 'primitive') summary of the relationship in question around which the points are scattered.

(b) It is perfectly legitimate to enter these values as 89, 90, 91, etc. This is equivalent to subtracting 1900 from the original numbers and doesn't alter the shape of the regression line but merely moves it to a different position on the graph. Thus, as long as the year axis is scaled '89, '90, '91, etc, it will look exactly the same.

(c) Changing the scales of either axis may alter the slope in terms of how steep it *looks* on the graph but, since the actual value for '*b*' remains unchanged, the effect is purely visual. The important general point to come out of this is that you shouldn't pay too much attention to the way the slope of the graph looks, as this is greatly affected by the scales used on the axes and whether or not there is a 'break' on the vertical axis. This and other related questions were discussed more fully in Chapter 6, *Lies and statistics*.

(d) The choice of which variable should go on which axis is usually *not* arbitrary. By convention, we try to choose the variable which is clearly not dependent on the other and place it on the horizontal axis. For example, it seems reasonable to assume that a person's age is independent of their pulse rates, rather than the other way around. So, in Figures 10.5 and 10.6, the variable 'Age' has been placed along the horizontal axis. However with some scattergraphs the choice may be more arbitrary, as with rainfall and sunshine, for example. It isn't obvious here which of these variables can be better thought of as being independent of the other. This decision and the whole question of dependence will be discussed in more detail in the next chapter.

(e) The most common types of data where regression is used are those where the variable 'time' is drawn on the horizontal axis – known as 'time series' graphs, for obvious reasons. Extrapolations made from time series graphs allow us to make backward or forward projections, which are interesting if we wish to make historical guesses about the past, or forecasts about such things as the population and the economy in the future.

(f) Clearly, if the scattergraph shows up a very clear linear pattern, then the best-fit line will be an extremely accurate summary of the relationship with all the points lying quite close to the best-fit line. In this situation predictions can be made with a good degree of confidence. However, if the points on the original scattergraph had no clear pattern, with a very wide scatter, then the line will have little predictive power. Just how confident you can be in the predictions that you make from the regression line will be discussed as the central theme of the next chapter.

Summary

This chapter dealt with how we might explore patterns in 'paired data'. Visually, the easiest way to see relationships between two variables is to plot the data onto a scattergraph. If the pattern of points appears to lie on a straight line, a useful way of summarizing the relationship is to find the equation of the 'line of best fit', or regression line, which can be made to pass through the scatter of points. Where the regression line has a positive slope, we say that the trend is positive – i.e. as one increases, so the other increases. Negative slopes suggest negative (sometimes known as 'inverse') relationships.

Regression lines are useful for making forecasts. Particular dangers are attendant where this involves extrapolation (extending the regression line beyond known data) since there is no guarantee that the observed pattern will continue to apply.

Regression can be thought of as a way of summarizing the relationship between two variables by collapsing the scatter of points into a single line. As such, the regression line reveals nothing at all about how widely scattered the points were around it. But what exactly does the degree of scatter tell us? This is an important question in statistics and is the central idea of the next chapter.

Calculating the regression coefficients using the formula

An accurate method for finding the regression coefficients is to perform a rather complicated calculation, using the formulas given below.

To find the slope, b, you first need to calculate the following intermediate values:

$n \rightarrow$ this is the number of pairs of values (in this example, $n = 12$)

$\Sigma XY \rightarrow$ the sum of the twelve XY products

$\Sigma X \rightarrow$ the sum of the X values

$\Sigma Y \rightarrow$ the sum of the Y values

$\Sigma X^2 \rightarrow$ the sum of the squares of the X values

These intermediate values are then substituted into the following exciting-looking formula (which will not be justified but simply stated here).

$$b = \frac{n\Sigma XY - \Sigma X\Sigma Y}{n\Sigma X^2 - (\Sigma X)^2}$$

Having calculated the value for b, finding a follows more easily. The mid-point of all the points on the scattergraph is denoted by the co-ordinates $(\bar{X}, \bar{Y})$. (*Note:* The sigma (Σ) and bar $(\bar{X}, \bar{Y})$ notations used above are explained in Chapter 2.)

An important property of the best-fit line is that it passes through this mid-point. That being so, we can substitute these co-ordinates, $(\bar{X}, \bar{Y})$, into the equation of the line $Y = a + bX$. This gives:

$$\bar{Y} = a + b\bar{X}$$

Rearranging this equation produces the following:

$$a = \bar{Y} - b\bar{X} \ldots \text{from which we can find the value of } a.$$

The example below uses this method to calculate the regression coefficients a and b based on the Rainfall/Sunshine data taken from Table 10.1. In order to simplify the calculation, we shall call inches of rainfall Y and hours of sunshine X.

First, let us find b. From the formula for b shown above, it will be necessary to find the following intermediate values: ΣXY, ΣX, ΣY and ΣX^2.

MONTH	HOURS OF SUNSHINE (X)	INCHES OF RAINFALL (Y)	X^2	XY
January	1.0	6.6	$(1.0)^2 = 1.00$	$1.0 \times 6.6 = 6.60$
February	1.8	4.1	$(1.8)^2 = 3.24$	$1.8 \times 4.1 = 7.38$
March	3.3	3.3	10.89	$3.3 \times 3.3 = 10.89$
April	4.3	3.2	18.49	$4.3 \times 3.2 = 13.76$
May	5.8	3.0	33.64	$5.8 \times 3.0 = 17.40$
June	6.1	3.1	37.21	$6.1 \times 3.1 = 18.91$
July	4.7	4.3	22.09	$4.7 \times 4.3 = 20.21$
August	5.3	5.0	18.49	$4.3 \times 5.0 = 21.50$
September	3.4	5.6	11.56	$3.4 \times 5.6 = 19.04$
October	2.3	6.8	5.29	$2.3 \times 6.8 = 15.64$
November	1.5	6.0	2.25	$1.5 \times 6.0 = 9.00$
December	0.9	5.9	0.81	$0.9 \times 5.9 = 5.32$
				$\Sigma XY = 165.64$
TOTALS	$\Sigma X = 39.4$	$\Sigma Y = 56.9$	$\Sigma X^2 = 164.96$	
$n = 12$, so:	$\bar{X} = \dfrac{39.4}{12}$	$\bar{Y} = \dfrac{56.9}{12}$		

Table 10.5 Calculating the regression coefficients

Knowing what we need to find, the next stage is to set up a table of values so that these intermediate values can be calculated systematically. A suitable table might look something like Table 10.5.

Note that when using a calculator it is not necessary to fill in each cell in the table above since it is only the totals we are interested in. Here is the formula for *b* again:

$$b = \frac{n\Sigma XY - \Sigma X \Sigma Y}{n\Sigma X^2 - (\Sigma X)^2}$$

By substitution from the table, we get:

$$b = \frac{12 \times 165.64 - 39.4 \times 56.9}{12 \times 164.96 - (39.4)^2}$$

$$= \frac{-254.18}{427.16}$$

$= -0.595$ (to three decimal places) $\rightarrow$ *Slope*.

Having found the value of *b*, we use it to find *a*:

Notice that $\bar{Y} = \dfrac{\Sigma Y}{n}$ and $\bar{X} = \dfrac{\Sigma X}{n}$

$a = \bar{Y} - b\bar{X}$

$\quad = \dfrac{\Sigma Y}{n} - b \times \dfrac{\Sigma X}{n}$

$\quad = \dfrac{56.9}{12} - b \times \dfrac{39.4}{12}$

$\quad = 4.742 - (-0.595) \times 3.283$

$\quad = 6.695$ (to three decimal places) $\rightarrow$ *Intercept.*

So, the regression equation is:

$$Y = 6.695 - 0.595X$$

Finally, it is worth reminding yourself what the variables, X and Y were used to represent.

$\quad X$ = Average number of hours of sunshine per day
and Y = Average number of inches of rainfall per month

So, 'Average inches of rain' = $6.695 - 0.595 \times$ 'Average hours of sunshine'.

What this means is that, if you know that the average number of hours of sunshine for a particular month in the Lake district was, say, 5 hours, then the predicted rainfall for that month could be estimated using the formula thus:

Average inches of rain $= 6.695 - 0.595 \times 5$
$\qquad\qquad\qquad\qquad = 6.695 - 2.975$
$\qquad\qquad\qquad\qquad = 3.72$

Comments on exercises

Exercise 10.2
Reading from the two bar charts gives the estimates shown in Table 10.6.
The 12 points are plotted as shown in Figure 10.8

MONTH	SUNSHINE (HOURS)	RAINFALL (INCHES)
Jan	1.0	6.6
Feb	1.8	4.1
Mar	3.3	3.3
April	4.3	3.2
May	5.8	3.0
June	6.1	3.1
July	4.7	4.3
Aug	4.3	5.0
Sept	3.4	5.6
Oct	2.3	6.8
Nov	1.5	6.0
Dec	0.9	5.9

Table 10.6   Sunshine and Rainfall data estimated from Figure 10.1

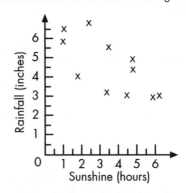

Figure 10.8   Scattergraph showing the data from Table 10.6

Exercise 10.4

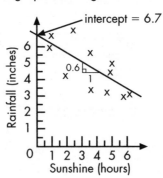

Figure 10.9   Best-fit line showing the relationship between rainfall and sunshine

(a) From Figure 10.9, the best-fit line cuts the Y axis at around 6.7 inches of rainfall. So: $a = 6.7$

(b) Figure 10.9 has focused on a small section of the best-fit line which is one unit wide. Over that interval of one unit in the X direction, the graph has fallen by around 0.6 units in the Y direction. Because this is a *fall*, it must be recorded as a *negative slope*. So: $b = -0.6$. (A more accurate approach might be to choose a wider interval of, say, 4 units in the X direction, measure the fall in the Y direction over this interval, and then divide by 4 to get the average fall per unit interval.)

(c) Substituting the values for a and b into the general equation, $Y = a + bX$, we get: $Y = 6.7 - 0.6X$

Exercise 10.5

(a) The scattergraph will look like Figure 10.10.

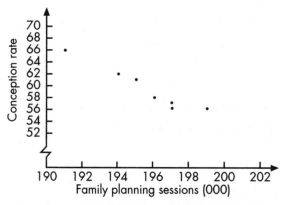

Figure 10.10 Scattergraph showing the relationship between the conception rate for women under 20 and the number of family planning clinic sessions

(b) The regression line, or 'line of best fit' has the equation:

$$Y = 224 - 0.84X$$

Table 10.7 illustrates the formula method for calculating the regression coefficients. The regression line has been

YEAR	FAMILY PLANNING CLINIC SESSIONS (X)	CONCEPTION RATE (Y)	X^2	XY
1	202	58	40 804	11 716
2	197	57	38 809	11 229
3	197	56	38 809	11 032
4	199	56	39 601	11 144
5	196	58	38 416	11 368
6	195	61	38 025	11 895
7	194	62	37 636	12 028
8	191	66	36 481	12 606
TOTALS	$\Sigma X = 1571$	$\Sigma Y = 474$	$\Sigma X^2 = 308\,581$	$\Sigma XY = 93\,018$
$n = 8,$	so: $\bar{X} = \dfrac{1571}{8}$	$\bar{Y} = \dfrac{474}{8}$		

Table 10.7 Calculating the regression coefficients for the family planning data

drawn on the graph and clearly does seem to fit the points well.

Here is the formula for b again:

$$b = \frac{n\Sigma XY - \Sigma X \Sigma Y}{n\Sigma X^2 - (\Sigma X)^2}$$

By substitution from the table into the formula, we get:

$$b = \frac{8 \times 93018 - 1571 \times 474}{8 \times 308581 - (1571)^2}$$

$$= \frac{-510}{607}$$

$$= -0.840 \text{ (to three decimal places)} \rightarrow Slope.$$

Having found the value of b, we use it to find a:

$$a = \bar{Y} - b\bar{X}$$

$$= \frac{\Sigma Y}{n} - b \times \frac{\Sigma X}{n}$$

$$= \frac{474}{8} - b \times \frac{1571}{8}$$

$= 59.25 - (-0.840) \times 196.375$

$= 224.244$ (to three decimal places) → *Intercept*.

So, the regression equation is:

$Y = 224.244 - 0.840X$

Where:

$X =$ Number of family planning sessions in England (000)

and $Y =$ Conception rate per 1000 women under 20, England & Wales

(c) The regression line has a negative slope. This means that there is a negative relationship between the two variables in question, namely that the rate of conceptions of women under the age of twenty has increased as family planning clinic sessions has fallen. However, there is no proof from this that the one has *caused* the other. Proving causation is hard and this important question will be looked at more fully in the next chapter.

Exercise 10.6

(a) The scattergraph and corresponding best-fit line are shown in Figure 10.11.

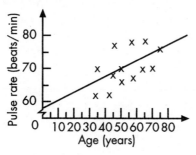

Figure 10.11 Scattergraph showing pulse rate and age

The trend shows a positive relationship between age and pulse rate – i.e. if the evidence of this sample is typical of the adult population as a whole, then in general, the older you are, the higher your pulse rate.

(b) The linear regression line calculated from these points is:

$Y = 58.24 + 0.23X$

where Y is pulse rate in beats per minute and X is age in years.

(c) Using the best-fit line calculated in (b), these values can be estimated by reading them off the graph, as follows:

To estimate Mia's pulse rate, start from her age (60) on the X axis draw a line vertically up to the best-fit line and finally draw across to the Y axis. This gives an estimate of Mia's pulse rate of around 72 beats per minute.

Mayerling's estimated pulse rate, by the same method, is around 58 beats per minute.

To estimate Tom's age, the same process is done in reverse. Start with the position on the Y axis which corresponds to Tom's pulse rate – 90 beats per minute – draw a line horizontally across until it meets the best-fit line, and then, from the X axis, read off the age which this corresponds to. In this case, the estimate of Tom's age = 138 years!

(d) We have made a dangerous assumption with these estimates, namely that the same linear trend will continue in both directions beyond the range of values in the sample. In fact, this is not valid in either case here, as you will see when you turn back to the text on page 172.

11 CORRELATION: MEASURING THE STRENGTH OF A RELATIONSHIP

A long drought in Damascus in 1933 occurred just after a national craze of playing with a yo-yo. In their wisdom, the city elders decided that the movement of the yo-yos was influencing the weather and promptly banned them. The next day ... it rained! Meanwhile, in Madrid, a bull-fighter, on the day of a *corrida*, will always shave his face twice. This is based on the belief that fear makes the beard grow faster and no bull-fighter wants to display fear in the ring.

Most of us have a few 'off the wall' beliefs of this nature. I know someone who refuses to travel by train because, about ten years ago he thinks he caught a cold in one! Based, often, on a tiny sample of experience or a few dubious anecdotes, we may deduce a cause-and-effect relationship where none exists.

This chapter builds on the ideas of regression which were the subject of the previous chapter and explores the degree of confidence we can have that the relationship under consideration really does exist. In other words, we are interested in knowing about the strength of the relationship. An intuitive sense of how strong a relationship is can be deduced just by looking at the data drawn in a scattergraph and this is where we start in the opening section of the chapter. The next two sections deal with formal measures of correlation, the product-moment correlation coefficient and the rank correlation coefficient. Finally we look at the difficult question of whether strong relationships are necessarily cause-and-effect ones.

Scatter

As has already been suggested in the previous chapter, the regression line is useful as a statement of the underlying trend but tells us nothing about

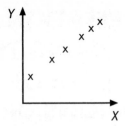

Figure 11.1   Scattergraph showing perfect positive correlation

how widely scattered the points are around it. The degree of scatter of the points, or the *correlation*, is a measure of the strength of association between two variables. For example, perfect positive correlation will look like Figure 11.1.

Perfect negative correlation will result from data which are plotted as in Figure 11.2.

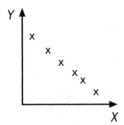

Figure 11.2   Scattergraph showing perfect negative correlation

In practice, most scattergraphs portray relationships with correlations somewhere between these two extremes. Some examples are shown in Figures 11.3 to 11.5.

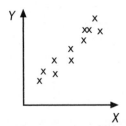

Figure 11.3   Strong positive correlation

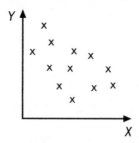

Figure 11.4 Weak negative correlation

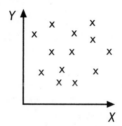

Figure 11.5 Zero correlation

EXERCISE 11.1. Racing pulses?

The two scattergraphs shown in Figures 11.6 and 11.7 below have been taken from Chapter 10 and both show the relationship between age and pulse rate for two different age groups.

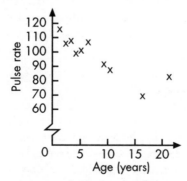

Figure 11.6 Scattergraph showing pulse rate and age of 10 young people – Sample A

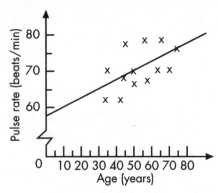

Figure 11.7 Scattergraph showing pulse rate and age of
14 adults – Sample B

a) *How would you describe these relationships?*
b) *How would you describe the strength of association in
each case and what might you deduce from this?*

Comments below

(a) The scattergraph for Sample A shows a clear negative relationship. Over the age range from 0 to 20 years, the pulse rate
typically drops from around 110 to about 65 beats per
minute. For the adults in Sample B, the relationship is a positive one, with the pulse rate rising from around 65 for young
adults to about 75 beats per minute for adults of around 70 to
80 years. We might deduce from this that children's pulse
rates *decline quite a lot* as they get older, but with adults there
is evidence of a *slight increase* in pulse rates with age.

(b) Because the points on the scattergraph for Sample A lie fairly
close to the best-fit line, there seems to be a strong negative
correlation here. The corresponding pattern in Figure 11.7
suggests a rather weaker, but positive, correlation for Sample
B. What this implies is that you can expect to be able to make
predictions about the link between pulse rate and age with
greater confidence for children than for adults.

It is worth stressing that parts (a) and (b) in the previous exercise deal
with different issues. Part (a) is concerned with describing *the basic*

underlying relationship as represented by the best-fit line and is therefore a question about regression. Part (b) directs attention to *the strength of the relationship* as represented by the degree of scatter of the points around the regression line and is therefore all about correlation.

Product-moment correlation coefficient

Although it is useful to be able to gauge the strength of the relationship simply by looking at the scattergraph, a more formal method based on calculation is available which gives a numerical value for the degree of correlation. The product-moment coefficient of correlation (sometimes known as Pearson's coefficient) is calculated on the basis of how far each point lies from the 'best-fit' regression line. It is denoted by the symbol r.

The formula for r, the product-moment correlation coefficient is:

$$r = \frac{S_{XY}}{S_X S_Y}$$

where X is the variable on the horizontal axis
and Y is the variable on the vertical axis

S_{XY} is the covariance, given by the formula $= \dfrac{\sum XY}{n} - \bar{X}\bar{Y}$

S_X is the standard deviation of $X = \sqrt{\dfrac{\sum X^2}{n} - \bar{X}^2}$ (1)

S_Y is the standard deviation of $Y = \sqrt{\dfrac{\sum Y^2}{n} - \bar{Y}^2}$

The formula has been devised to ensure that the value of r will lie in the range -1 to 1. For example:

$r = -1$ means perfect negative correlation
$r = 1$ means perfect positive correlation
$r = 0$ means zero correlation
$r = -0.84$ means strong negative correlation
$r = 0.15$ means weak positive correlation

and so on.

[1] This is a slightly different version of the formula for standard deviation given in Chapter 5, but they both amount to the same thing.

Certain calculators with specialist statistical functions provide a key (usually marked 'r') for directly calculating the correlation coefficient. The exact procedure will be explained in the calculator manual but in principle you must first enter all the pairs of data values, then follow by a press of the 'r' key. With only a conventional calculator to hand, however, calculating the value of r is really quite an effort. The following simple example sets out the three main steps in which the covariance, S_{XY}, the standard deviation of the X values, S_X, and the standard deviation of the Y values, S_Y, are first calculated.

A simple worked example for calculating r

Let us suppose that you have a simple data set consisting of only the following three pairs of values

X	Y
1	2
2	3
3	4

Step 1   *Calculating the Covariance,* S_{XY}

X	Y	XY
1	2	2
2	3	6
3	4	12
$\sum X = 6$	$\sum Y = 9$	$\sum XY = 20$

$$\bar{X} = \frac{6}{3} = 2 \quad \bar{Y} = \frac{9}{3} = 3$$

$$S_{XY} = \frac{20}{3} - 2 \times 3$$

$$= 6\frac{2}{3} - 6 = \frac{2}{3}$$

Step 2   *Calculating the standard deviation of* X, S_X

X	X^2
1	1
2	4
3	9
$\sum X = 6$	$\sum X^2 = 14$

$$\bar{X} = 2$$

$$S_X = \sqrt{\frac{14}{3} - (2)^2}$$

$$= \sqrt{\frac{2}{3}}$$

Step 3   *Calculating the standard deviation of* Y, S_Y

Y	Y^2
2	4
3	9
4	16

$\sum Y = 9$ $\sum Y^2 = 29$

$\bar{Y} = 3$

$$S_Y = \sqrt{\frac{29}{3} - (3)^2}$$

$$= \sqrt{\frac{2}{3}}$$

Finally, $r = \dfrac{S_{XY}}{S_X S_Y} = \dfrac{\dfrac{2}{3}}{\sqrt{\dfrac{2}{3}} \times \sqrt{\dfrac{2}{3}}} = \dfrac{\dfrac{2}{3}}{\dfrac{2}{3}} = 1$

So in this example there is perfect positive correlation. When the three points are plotted on a scattergraph, this should make sense just by looking at their pattern. As you can see from Figure 11.8, the points form a perfect straight line with a positive slope.

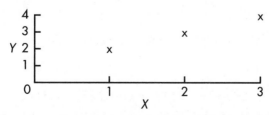

Figure 11.8   Data from simple worked example showing perfect positive correlation

EXERCISE 11.2. Another simple example showing the calculation of r

You can try an example on your own now. At this stage, you may still find it helpful to use simple artificial data just to clarify the main steps. For example, try the data below, which show perfect negative correlation. So you should expect to get a value for $r = -1$.

X	Y
1	4
2	3
3	2

Worked example

The next worked example is based on rather more interesting data. Notice how Table 11.1 provides you with an efficient, systematic way of calculating S_{XY}, S_X and S_Y. Here the calculation for r is based on the data given in Sample B in Exercise 10.8.

X	Y	X^2	Y^2	XY
(age)	(pulse rate)			
33	62	1089	3844	2046
50	66	etc	etc	etc
45	77			
65	78			
57	78			
70	70			
42	62			
75	76			
57	67			
35	70			
44	68			
63	70			
50	70			
55	72			
$\Sigma X = 741$	$\Sigma Y = 986$	$\Sigma X^2 = 41\,281$	$\Sigma Y^2 = 69\,814$	$\Sigma XY = 52\,662$
$\bar{X} = 52.93$	$\bar{Y} = 70.43$			

Table 11.1 Pulse rate and age of 14 adults (n = 14)

$$S_{XY} = \frac{52\ 662}{14} - 52.93 \times 70.43 = 33.71$$

$$S_X = \sqrt{\frac{41\ 281}{14} - (52.93)^2} = 12.127$$

$$S_Y = \sqrt{\frac{69\ 814}{14} - (70.43)^2} = 5.131$$

Coefficient of correlation, $r = \dfrac{33.71}{12.127 \times 5.131} = 0.542$

Clearly this calculation requires considerable effort, even with the aid of a calculator. If you need to find r repeatedly, it would be worth investing in a 'statistical' calculator which calculates it directly from the raw data.

EXERCISE 11.3. Over to you!

In order to get practice at calculating the product-moment coefficient of correlation, have a go at some or all of the following and check that you come up with the same results as shown.

NAME	AGE (YRS)	PULSE RATE (BEATS PER MINUTE)
Irena	10	86
Nadia	3	107
Luke	4	98
Jon	6	105
Carrie	8	90
Theo	1	115
Jagjeet	16	69
Ira	21	81
Hannah	5	100
Alex	2	105

Table 11.2   *The ages and pulse rates of a sample of 10 young people*
Source: Personal survey

Comments follow

The correlation coefficient for age against pulse rate for Sample A is -0.88. Intermediate values for calculating this value are given below.

$$S_{XY} = \frac{6553}{10} - 7.6 \times 95.6 \quad = -71.26$$

$$S_X = \sqrt{\frac{952}{10} - (7.6)^2} \quad = 6.12$$

$$S_Y = \sqrt{\frac{93\,146}{10} - (95.6)^2} \quad = 13.24$$

Coefficient of correlation, $r = \dfrac{-71.26}{6.12 \times 13.24} = -0.89$

EXERCISE 11.4. Interpreting r

How do you interpret the two values for r given in Exercise 11.3 and the worked example above?

Comments below

The strength of correlation will be indicated by how far from zero the value of r lies. The Sample B value of $r = 0.542$ is reasonably strong, but the Sample A value of -0.88 is very high indeed, and suggests that among children there is a strong (negative) correlation between the two variables concerned. However, there are two important provisos which need to be considered when interpreting correlation.

The first point is the question of *batch size*. For Sample A, the batch size is 10 and for Sample B it is 14. In general, the smaller the batch size, the larger must be the value of r in order to provide evidence of significant correlation. Thus, we must not be overly impressed with the Sample B value for r of -0.88 since the batch size is so small. This point, which relates to the degree of *significance* which we can attach to values of r is explained more fully in the next chapter.

Secondly, a crucial issue here is to be able to distinguish correlation between two variables and a *cause-and-effect relationship* between them. It is important to remember that correlation only suggests a pattern linking two sets of data. CORRELATION DOES NOT PROVE THAT ONE OF THE VARIABLES HAS *CAUSED* THE OTHER. Indeed, exactly what the value of r does and does not tell us is quite difficult to interpret, and this issue is pursued in more detail in the final section of this chapter.

Rank correlation

It is not always necessary, or even possible, when investigating correlation, to draw on the sort of measured data that have been presented so far. An alternative is to work only from the *rank* positions. For example, when judging ice-skating or gymnastics competitions, the marks that are awarded are rather meaningless and their primary purpose is to allow the competitors to be ranked in order – first, second, third, and so on. Information is sometimes stripped of the original data and the results presented only in order of rankings. For example, look at Table 11.3 which shows the top 10 boys' and girls' first names of 1991 published in *The Times*. What, if anything, do these results suggest about the names that tend to get chosen?

The first point to be made here is that these choices may say something about the names preferred by readers of *The Times* but say little about the preferences of readers of, say, the *Daily Mail* or the *Daily Sport*. However, there are some interesting investigations that can be made about the different patterns between girls' and boys' names. For

		BOYS			GIRLS
1	[1]	James	1	[4]	Emily
2	[3]	Alexander	2	[–]	Katherine
3	[2]	Thomas	3	[1]	Charlotte
4	[4]	William	4	[9]	Olivia
5	[6]	Charles	5	[2]	Sophie
6	[7]	Edward	6	[3]	Lucy
7	[5]	Oliver	7	[–]	Hannah
8	[8]	George	8	[6]	Alice
9	[–]	Matthew	9	[10]	Georgina
10	[9]	Henry	10	[7]	Emma

Table 11.3 The top 10 first names of 1991 (1990 rankings in brackets)
Source: The Sunday Times, 29 December 1991

example, notice here how many of the girls' list contains names ending in a vowel (7) as compared with the number of boys' names (1). You may also have observed that the boys' list appears to be rather traditional, but quite how this characteristic could be measured and compared is not easy to see.

One property that we *can* test is the extent to which the two sets of names have retained their popularity since the previous year. A simple way to get a quick visual impression of this is to plot the rank positions of one year against the other on a scattergraph. An immediate problem is how to rank the names for which last year's position is marked as [–], i.e. where the rank position was outside the top ten in the previous year. A simple solution is to give this name the next highest ranking available in the list. Thus, for Matthew the [–] is replaced by a rank of 10 (one below Henry's ranking for last year), while Katherine and Hannah are both given 11 (one above Georgina's position for last year). The two scattergraphs then look like Figures 11.9 and 11.10.

EXERCISE 11.5. Interpreting the scattergraphs
What do these two scattergraphs reveal?

Comments below

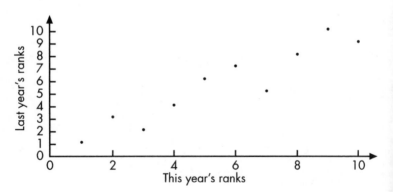

Figure 11.9 Scattergraph showing the 1991 ranks of boys' names against the previous year's position

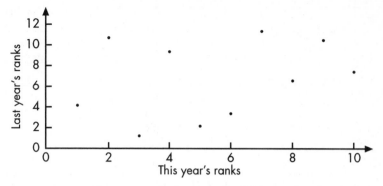

Figure 11.10   Scattergraph showing the 1991 ranks of girls'
names against the previous year's position

A clear pattern in the boys' scattergraph suggests a strong correlation
between the rank order of names in 1990 and 1991. There is little evidence of any pattern in the girls' names, revealing that the preferences for
boys' names changed less in popularity than the girls' names between the
two years in question. But can we demonstrate this difference quantita-
tively?

The main theme of this section is to try to find a way of *measuring* the
strength of association in these and similar scattergraphs where the data
are in ranked form. While it would be possible to use the correlation
coefficient, r, in such cases, there is an alternative measure which is specially designed for ranked data called the rank coefficient of correlation,
r'. It is sometimes known as 'Spearman's coefficient of rank correlation'
after its inventor, Charles Spearman.

The formula for r' is based on calculating all the differences (d) between
each pair of ranks.

The sum of the
squares of the
differences in
ranks

Coefficient of rank correlation, $r' = 1 - \dfrac{6\sum d^2}{n(n^2 - 1)}$

n is the batch size

BOYS	1991	1990	d	d^2
James	1	1	0	0
Alexander	2	3	−1	1
Thomas	3	2	1	1
William	4	4	0	0
Charles	5	6	−1	1
Edward	6	7	−1	1
Oliver	7	5	2	4
George	8	8	0	0
Matthew	9	10	−1	1
Henry	10	9	1	1

Total = 10

Table 11.4   Calculating r′ for the boys' names

Here is a worked example using the boys' names data from Table 11.4.

$$r' = 1 - \frac{6 \times 10}{10(99)} = 0.939$$

As with the product-moment correlation coefficient, r, the coefficient of rank correlation, r', has been devised to ensure that its value will lie within the range −1 to 1. The value for r' can be interpreted in much the same way as the values for r were in the previous section. Thus:

> $r' = -1$ means perfect negative correlation
> $r' = 1$ means perfect positive correlation
> $r' = 0$ means zero correlation
> $r' = -0.84$ means strong negative correlation
> $r' = 0.15$ means weak positive correlation

and so on.

So clearly our value of 0.939 for the rank correlation coefficient for the boys' names shows a very high degree of correlation indeed. This is a statistical way of saying that the rank order of popularity of boys' names in 1991 was very similar to the rank order of the previous year.

EXERCISE 11.6. Calculating *r'* yourself

You will find it a useful exercise now to calculate a coefficient of rank correlation for yourself. Complete Table 11.5 and then work out *r'* for the girls' names data from the table. How does the result compare with the corresponding figure for boys, and what does the comparison mean in terms of the question we have posed?

GIRLS	1991	1990	d (1991–1990)	d^2
Emily	1	4		
Katherine	2	11		
Charlotte	3	1		
Olivia	4	9		
Sophie	5	2		
Lucy	6	3		
Hannah	7	11		
Alice	8	6		
Georgina	9	10		
Emma	10	7		

Total =

Table 11.5   Calculating r' for the girls' names

Comments below

You should have found that the value for r' for the girls' names was -0.012. The calculation is set out in full in Table 11.6.

For the girls' names, $r' = 1 - \dfrac{6 \times 167}{10(99)} = -0.012$

This value for r' is very close indeed to zero and suggests that there is no correlation between the girls' names chosen in 1990 and 1991. This is,

GIRLS	1991	1990	d (1991–1990)	d^2
Emily	1	4	−3	9
Katherine	2	11	−9	81
Charlotte	3	1	2	4
Olivia	4	9	−5	25
Sophie	5	2	3	9
Lucy	6	3	3	9
Hannah	7	11	−4	16
Alice	8	6	2	4
Georgina	9	10	−1	1
Emma	10	7	3	9

Total = 167

Table 11.6

however, a rather misleading conclusion to draw. It must be remembered that there were many more than ten girls' names in the overall rankings and of the top ten in 1991, as many as eight had been in the 1990 top ten. So, if we were to calculate the rank correlation for, say, the top 50 girls' names, a much more positive correlation might result. This raises an important general point in statistics, namely that you must be very careful, when attempting to draw general conclusions, that you don't over-generalize beyond the scope of the data on which the conclusions were based. In this particular example, we need to stress that we are identifying patterns in the top ten names chosen by readers of *The Times* and over a particular two year period in history. These conclusions are not valid for other situations.

Cause and effect

Actually, the less I practise, the better I seem to get.

The more I drink, the quicker my reactions are.

For certain everyday events, the link between cause and effect seems fairly straightforward. We turn on a tap and water comes out; we fall over and it hurts! You would have to be a Cartesian philosopher to doubt the obvious connections between action and outcome in these examples. But not all relationships are so easy to interpret. Suppose, for example, that you suffer from persistent back pain and have tried a variety of remedies to cure it. You waken one morning and think that it feels a bit better. Here is the sort of conversation you might be having with yourself.

Hmm ... it must be those exercises that I did last week are beginning to have an effect. Perhaps I should do these exercises more regularly. Of course, it could be due to the restful bubble bath that I took last night. That's worth remembering – (thinks 'I must buy another bottle of that bath oil ... '). Hang on, though! On second thoughts, isn't there a bit of pain shooting down the old left leg? Yes, it's definitely nagging there. I must have overdone it on the exercise machine. I'd better lay off a bit in future ...

The problem with back pains and their remedies is that these things are extremely difficult to measure. Also, there are so many possible explanations for the various outcomes, you can never be sure why they happened. As with many ailments, they may just get better over time anyway despite your attempts at a drastic cure.

With proper medication the common cold usually lasts about a week but left to its own devices it can drag on for seven days.

The relevance of these examples to correlation is to remind you that a strong correlation between two things does not prove that one has *caused* the other. A strong correlation merely indicates a statistical link, but there may be many reasons for this besides a cause and effect relationship. For example, traditionally the height of a pregnant woman and her shoe size were thought to be good predictors of whether or not she would need to give birth by Caesarean section operation. In general, it was thought, women who were short in stature and petite of foot would also have narrow pelvises and would therefore find it difficult to have a normal vaginal delivery. And indeed, the statistical evidence seems to bear this out, in that there has traditionally been a close correlation between women having a Caesarean birth and their having small feet. But this is not the same thing as saying that there is a close correlation between women *needing* a Caesarean birth and having small feet. It now seems that this is something of a myth which has grown up over time. Doctors, believing the association to be valid, tend to prescribe Caesarean births for women with small feet, thereby making the correlation a self-fulfilling prophecy.

Such an association between two things is known as *spurious correlation* because it implies a cause and effect link which is not actually there. There are countless further examples of this phenomenon. For example, the coefficient of correlation between population in 1989 and gross domestic product (GDP) in each of the 12 EU countries, when calculated gives:

$$r = 0.948$$

This is a very high value, indicating a strong link between the two variables. But does this suggest that increases in population *cause* an increase in wealth? Or is the relationship the other way round – i.e. that countries, as they become richer, can afford to support a larger population? Merely calculating a value for r does not begin to provide answers to these sorts of questions.

Here are some further examples. You might like to entertain yourself by trying to think up some different possible explanations for the connections.

EXERCISE 11.7. Causal or casual?

Try to find at least two explanations for the following.

  a) *The number of storks nesting in the chimneys of a sample of German villages is strongly correlated with the number of children in these villages.*
  b) *The number of cigarettes which people smoke is negatively correlated with their income.*
  c) *The number of crimes over time is positively correlated with the size of the police force.*

Comments below

  (a) *Babies and storks.* One possibility is that the storks are bringing the children to the villages. (Well, I suppose it is *possible*!) Another is that villages with more children tend to be larger and have more houses which therefore offer a larger number of chimneys for the storks to roost in.

  (b) *Smoking and wealth.* Do people smoke because they are poor or are they poor because they smoke?

  (c) *Police and crime.* Do we tend to employ greater numbers of police as a result of increases in crime, or do more

police simply catch more criminals? A third possibility may be that a larger police force encourages a higher proportion of victims to report crimes (number of crimes committed is not the same thing as number of crimes reported). It is also likely that changes in patterns of crime over time may have less to do with the quality and quantity of the police officers we employ and much more to do with changes in the economic and value system of society as a whole.

These examples illustrate the difficulties in interpreting the result of being told that there is a strong correlation between two things. We really can't be sure which one was the cause and which the effect, or indeed whether both things were quite independent of each other and the changes were because of some other factor or factors.

Cause and effect can really only be tested under controlled conditions where all other possible influences are removed. Then one factor (the independent one) is systematically altered and the resulting change on the other one (the dependent factor) is observed. This is discussed in more detail in the section headed 'Controlled trials' in the final chapter.

EXERCISE 11.8. Lollipops or rocket launchers?
a) *Using the data given in Table 11.7, calculate the coefficient of correlation between girls' average weekly pocket money over the period 1986 and 1990 and the number of offenders cautioned for violent offences against the person over the same period.*
b) *What explanation can you come up with for the resulting strong correlation?*

YEAR	GIRLS' AVERAGE WEEKLY POCKET MONEY (PENCE)	CAUTIONS FOR VIOLENT OFFENCES (THOUSANDS)
1986	114	9.5
1987	120	11.3
1988	122	12.7
1989	136	14.7
1990	147	16.8

Table 11.7   Girls' average weekly pocket money and cautions for violent offences against the person in England and Wales
Source: Walls Pocket Money Monitor and Social Trends 22

Comments follow

The value for r for these data comes to approximately 0.98. However, this high level of correlation does not necessarily imply that schoolgirls are actually buying arms with their extra pocket money and supplying them to the underworld. What is more likely is that pocket money, like many things, has risen steadily with inflation and, for reasons which are not easy to establish, violent crime has also risen steadily. A simple cause and effect relationship is possible in this situation, but unlikely. A final point here, is that this value for r of 0.98 was calculated on the basis of only 5 pairs of values – much too small a batch size to base any real significance on the figure.

Politicians in office will point to a lowering of inflation and a rise in investment in the economy and expect to take all the credit, claiming that these were directly caused by their policies. Politicians in opposition will point to a lowering of educational standards and a rise in unemployment and expect the government to take the blame, claiming that these were directly caused by their policies. Political 'hawks' in the early 1990s claimed that an easing in world tension (the ending of the 'cold war') happened *because* they took a hard line and talked tough. The 'doves' argued that these changes happened *despite* the tough talking of the hard-liners and would all have taken place anyway for a set of quite different reasons.

Real world judgements are not made in controlled laboratory conditions and the true nature of cause and effect relationships will always be open to many interpretations, subject to individual opinions, prejudices and vested interests.

Summary

Correlation is a measure of the strength of association between two things. A useful intuitive indication of correlation is the extent of the scatter around the best-fit line when the data are plotted on a scattergraph. Positive correlation shows a clear pattern of points running from bottom left to top right, while negative correlation shows a pattern of points running from top left to bottom right.

Two measures of correlation were introduced: the Pearson coefficient (r) and, for ranked data, the Spearman coefficient (r'). Both measures produce values lying in the range -1 to 1. A coefficient value of -1 means perfect negative correlation, while a coefficient value of 1 means perfect positive correlation. When r or r' take a value of around zero, there is no correlation between the two factors in question.

Finally, caution was urged in the interpretation of correlation. Strong correlation is no indication of whether or not the relationship in question is one of cause and effect.

12 CHANCE AND PROBABILITY

Probability is a way of describing the likelihood of certain events taking place. Clearly different events have different likelihoods of occurring. At one extreme, some events may be thought of as being *certain* (the chance that you will eventually die, for example, is a certainty for mortal readers of this book), while at the other extreme another event might be described as *impossible* ('flying pigs', maybe). But for most everyday situations, the degree of likelihood involved is neither certain nor impossible but will lie somewhere between these two extremes. You might like to spend a few minutes thinking about events whose outcomes are uncertain and some of the words we use to describe the degree of uncertainty.

You can talk about chance in terms of being 'very likely', 'fairly common', 'extremely unlikely', 'an even chance', 'a long shot', 'beyond all reasonable doubt', and so on. Although these words are in common use, there is considerable empirical evidence to suggest that they don't mean the same thing to everyone. For example, some people would rate 'fairly common' as having a greater than even chance, while others would rate it much lower.

For most everyday purposes, these differences in meaning may not matter very much. However, there are other situations where words alone are just not sufficiently precise to describe and compare probabilities accurately.

Measuring probability
Odds

Odds are the most common way of measuring uncertainty in situations where people are betting on an event whose outcome is unknown. In a horse race, for instance, the odds on each horse are a measure of how likely the bookmaker thinks each horse is to win a particular race. For example, if a horse is given odds of 5–1 (said as 'five to one' and usually written either as 5–1 or 5:1) this means that, given six chances, it would be expected to win once and lose the other five times. Suppose you bet on a horse with odds of 10:1 and the horse wins, then for each £1 you bet, you win £10. If, instead, you had placed a bet of £5 you would win £50 as well as receiving the £5 that you had staked in the first place. If a horse is favourite to win, it is given short odds, perhaps 2:1 or 3:2; whereas unlikely winners will be given very long odds, perhaps 100:1. A near certainty might be described in betting jargon as an 'odds-on-favourite'. This is where a horse is given such short odds that a winning bet will earn the punter less than their stake. Of course, they will get their stake back also. An example of an odds-on favourite might be odds of '2 to 1 on'. These are actually odds of '1 to 2'. In other words, betting £2 wins you a mere £1. The next exercise will give you a chance to calculate betting odds.

EXERCISE 12.1. Calculating odds

Below are the starting prices for an eight-horse race as they would appear in a newspaper. The betting is given at the bottom.

1	011	PLAY THE ACE (40) (E R Thomas) J Berry 9 7 .... J Carroll 2
2	2525	AZUREUS (IRE) (16) (J C Murdoch) J S Wilson 9 5................... J Fanning (7) 5
3	414641	LAND SUN (IRE) (16) (John W Mitchell) M Channon 9 3 .... C Rutter 8 V
4	1	COLWAY DOMINION (24) (K Stringer) J Watts 9 3................. Dean McKeown 7
5	153266	STAMFORD BRIDGE (17) (Mel Brittain) M Brittain 9 2................... M Wigham 6
6	1	MARTINI EXECUTIVE (9) (D) (Martini Cars (Cheshunt Ltd) W Pearce 9 1...... D Nicholls 4
7	054	AMANDHLA (IRE) (75) (Neill Jackson) N Tinkler 9 0..................... K Darley 1
8	364	FYAS (25) (M H Easterby) M H Easterby 7 11 .... L Charnock 3

—8 declared—

BETTING: 9-4 Martini Executive, 3-1 Colway Dominion, 5-1 Land Sun, 7-1 Play the Ace, 10-1 Azureus, 12-1 Stamford Bridge, 14-1 Fyas, 16-1 Amandhla

1989: Piquant 2 9 4 D McKeown 6-4 (W Hastings-Bass) 10 ran

Figure 12.1 The starting prices of an eight-horse race

Assuming you place a bet of £10 and ignoring the complications of tax, how much would you expect to win if you placed a winning bet on (a) the favourite, Martini Executive; (b) Amandhla; (c) Land Sun?

Comments on page 226

Statistical probability

In statistical work, probabilities are usually measured as numbers between zero and 1 and can be expressed either as fractions or as decimals. A probability of zero ($p = 0$) means zero likelihood, i.e. the outcome is impossible. A probability of one ($p = 1$) refers to an outcome of certainty. Most events have outcomes whose probabilities lie somewhere between zero and one. Thus, an event thought of as being fairly unlikely might have a probability of something like 0.1, while an 'even chance' would correspond to a probability of 0.5, and so on.

Probability is about calculating or estimating the likelihood of various outcomes from particular events. Clearly, if there is only one possible outcome from a particular event, it is certain to happen and therefore its probability will equal 1. So if we are involved in calculating probabilities, we must be dealing with events with more than one possible outcome. Table 12.1 gives a few examples. As you look at it, focus on

EVENT	POSSIBLE OUTCOMES	
Tossing a coin	1	Getting heads
	2	Getting tails
Choosing a playing card	1	Getting a spade
	2	Getting a heart
	3	Getting a diamond
	4	Getting a club
Having a baby	1	Getting a girl
	2	Getting a boy
Taking an examination	1	Passing
	2	Failing

Table 12.1   Events and outcomes

the distinction between an *event* and the *outcome(s)* of that event, as these two terms are often confused with each other. Understanding the distinction between these two terms will be particularly important when you study the sections describing mutually exclusive outcomes and independence later in this chapter.

You might like to add some of your own events and their corresponding possible outcomes. With some events, such as tossing a coin for example, there are only two possible outcomes.[1] With other events, there are literally hundreds – for example, when someone puts their hand in a bag to select the winning raffle ticket. Of course, from the punter's point of view, there are only two outcomes of any importance here – winning or losing! Sometimes outcomes are all equally likely for a given event. For example, if you select a card from a deck of 52 playing cards which has been thoroughly shuffled, it is equally likely that the card you choose is a spade, heart, diamond or club. The reason for this is that there are equal numbers of each suit – 13 in fact. With other events, the various outcomes are not equally likely. For example, with most examinations, more people tend to pass than fail and therefore the probability of a randomly chosen individual passing is more likely than their failing. Similarly, slightly fewer girls are born than boys, so the probability of a baby being a girl is lower than being a boy (about 0.48 as against 0.52).

This would be a good moment to reflect briefly on these two words, 'event' and 'outcome'. Unfortunately the clear distinction that has been made here between events and outcomes is not universal in statistics textbooks. Many authors use the term 'event' when they are actually describing an 'outcome'. In some more mathematical textbooks the distinction is made with the words 'experiment' and 'outcome', respectively. However, I do not favour the word 'experiment' as the term implies that only a narrow range of coin- and dice-tossing activities is being considered.

When outcomes are equally likely it is fairly straightforward to calculate their probabilities. You simply divide the number of ways in which that outcome can occur by the total number of possibilities. Some examples are given in Table 12.2. Now do Exercise 12.2.

In summary, a general definition of probability is $p = \dfrac{n}{N}$

where p is the probability of a particular outcome occurring, n is the number of ways that outcome can occur and N is the total number of possibilities.

It is important to remember that this definition of probability only applies in situations where the various ways in which the outcome in question

---

[1] There is a third outcome – that the coin lands on its edge. However, if the coin is tossed on to a smooth, hard surface, this outcome is so unlikely that its probability can be taken equal to zero.

EXERCISE 12.2. Events, outcomes and probability

Table 12.2 shows how probabilities of outcomes are usually calculated. Check that you understand the first two and fill in the rest for yourself.

EVENT	OUTCOME	NUMBER OF WAYS (n)	NUMBER OF POSSIBILITIES (N)	PROBABILITY $\left(\dfrac{n}{N}\right)$
Tossing a coin	Getting 'tails'	1	2	$\dfrac{1}{2}$
Tossing a die	Getting a score of 5 or more	2	6	$\dfrac{2}{6} = \dfrac{1}{3}$
Choosing a playing card	The suit being 'hearts'			
Choosing a playing card	The card being an 'ace'			
A baby being born ...	... on a weekend			
Someone who was born in December ...	... having their birthday *after* Christmas Day			

Table 12.2 Calculating probabilities

Comments on page 227

can occur are equally likely. Outside the cosy world of perfectly shuffled packs of cards and perfectly symmetrical dice, this assumption is often unrealistic. Thus, to take the two examples of a December baby being born after Christmas Day and a baby being born on a weekend, at first sight it may seem reasonable to assume that babies are equally likely to be born on all days of the year. Yet, for hospital births, this may not be so. Many babies come into the world as a result of being medically induced and, given that hospitals may be short-staffed at weekends and over the Christmas period, inductions may not be so popular at this time. This is one reason that most textbooks stick to 'safe' examples involving dice, coins and cards when dealing with probability concepts.

Myths and misconceptions about probability

There are many examples of faulty intuitions in the area of probability. Indeed it is quite possible that you have some yourself. Exercise 12.3 will give you an opportunity to test out your intuitions in this area.

EXERCISE 12.3. Beliefs and intuitions
Which of the following do you think are true?

   i) *If you throw a six-faced die, a 'six' is the most difficult outcome to get.*
   ii) *If a coin is thrown six times, three heads and three tails is the most likely outcome.*
   iii) *If a coin is thrown six times, of the two possible outcomes below, the second is more likely than the first (H = Heads, T = Tails)*
   Outcome A   H H H H H H
   Outcome B   H T T H H T
   iv) *If two dice are thrown and the results added together, scores of 12 and 7 are equally likely.*
   v) *If a fair coin was tossed ten times and each time it showed 'heads' you would expect the eleventh toss to show 'tails'.*

Comments below

   (i)   Although many people believe that a six is the most difficult score to get when a die is tossed, this is actually false. There are several possible explanations for why this mistaken notion might persist. One may be that six is the biggest score, and therefore thought to be hardest to throw. A second may be that in many dice games, six is

the score that is required and therefore it seems logical that it should be the most difficult to get. But perhaps the most likely explanation is that people correctly feel that it is less likely to toss a six than to toss a 'not-six', and therefore it is more difficult to get a six, than not to get a six. However, this is quite a different belief from that of thinking that the six outcomes on a die are not equally likely.

(ii) This statement is correct. There are seven possible outcomes for this event, which range from getting no heads, to getting six heads. The two least likely outcomes are the extremes of 'all six heads' and 'all six tails', whereas three heads and three tails is the most likely. The bar graph, Figure 12.2, summarizes the likelihoods of the various outcomes.

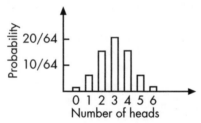

Figure 12.2 Bar chart of the likelihoods of the possible outcomes if a coin is thrown six times

Just why the graph looks like this and how to calculate the separate probabilities of each outcome (indicated on the vertical scale) are explained in the next chapter in the section called 'The binomial distribution'.

(iii) This statement is false. Although an outcome with three heads and three tails is more likely than an outcome with six heads, the two sequences given here are in fact equally likely. The reason for this is that there is only one way of getting six heads in a row (HHHHHH), whereas there are many ways of getting three heads and three tails – for example, HTHTHT or HHHTTT or TTTHHH etc.

So, in fact, these two sequences are equally likely. Again, this will be explained more fully in the next chapter.

(iv) This statement is false. There is only one way to throw a 'twelve' (with a double six), whereas there are many ways in which a 'seven' can be obtained. The bar graph, Figure 12.3, shows the likelihood of each of the eleven outcomes (2, 3, . . . 12) when two dice are thrown.

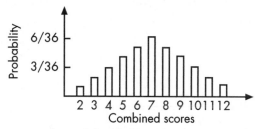

Figure 12.3 Bar chart of the likelihoods of the possible outcomes if two dice are thrown

(v) The final statement is also false. Many people believe that, according to some vague sense of the ill-defined 'law of averages', an unexpectedly long run of heads will necessarily be corrected for by a tail on the next throw. Whether it is the coin itself or God who is responsible for this even-handed averaging out of outcomes is unclear. The reality is that the coin has no memory of past outcomes and begins each new toss afresh. This misconception is sometimes known as the 'gambler's fallacy' as it leads punters to believe that, after a string of losses, their next bet is more likely to win.

Mutually exclusive outcomes and adding probabilities

When two (or more) outcomes are said to be 'mutually exclusive' this means that if one occurs, the other cannot occur. For example, if a coin is tossed, the two possible outcomes 'getting heads' and 'getting tails' are mutually exclusive, because both cannot occur at the same time. Not all events produce outcomes which are mutually exclusive. For example, when a baby is born, one outcome could be that it is a girl (rather than a boy). Another outcome might be that it has brown eyes (rather than blue or green). However, if a brown-eyed girl is born, both outcomes have occurred at the same time – clearly not mutually exclusive outcomes.

EXERCISE 12.4. Are the outcomes mutually exclusive?

Complete the table below, indicating whether you think the outcomes listed for the particular events are mutually exclusive.

EVENT	OUTCOMES	OUTCOMES MUTUALLY EXCLUSIVE? Y/N
(a) Tossing a coin	1 Getting heads 2 Getting tails	
(b) Tossing a die	1 Getting an even score 2 Getting an odd score	
(c) Tossing a die	1 Getting an even score 2 Getting a score greater than 3	
(d) Selecting a playing card	1 Getting an ace 2 Getting a spade	

Table 12.3   Events and outcomes

Comments on page 227

Have a closer look at event (c) in the table above. If the outcomes are represented as shown below, it is clear that 4 and 6 appear in both lists.

Even numbers	2 4 6
Numbers greater than 3	4 5 6

An alternative way of representing the outcomes of events is by using rings corresponding to the different possible outcomes. The two outcomes being considered are shown here as two overlapping rings – outcome E (getting an even score) and outcome G (getting a score greater than 3).

All possible outcomes

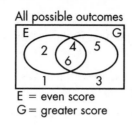

E = even score
G = greater score

Figure 12.4 Tossing a die

When outcomes are not mutually exclusive, as is the case here, the rings will overlap. The region where they intersect corresponds to the particular scores (4 and 6) where both outcomes occur at the same time.

You may be wondering why it might be useful to know whether outcomes of certain events are mutually exclusive. It is particularly important when *adding probabilities* to establish whether or not probabilities 'overlap', so as to avoid double-counting the region of intersection. A simple example based on outcomes E and G above should make this clearer.

The probability of outcome E can be written as *P(E)*. Similarly, the probability of outcome G can be written as *P(G)*.

Now $P(E) = 3/6 = 0.5$ (i.e. this is the chance of getting an even score)

and $P(G) = 3/6 = 0.5$ (i.e. this is the chance of getting a score greater than 3)

Supposing we were interested in finding the probability of either getting an even score or getting a score greater than 3, it would be tempting simply to add these two separate probabilities: $0.5 + 0.5$, giving the answer 1. However, this answer would seem to suggest that the probability of either one or other of *E* or *G* occurring is a certainty. This is clearly incorrect, because tossing a one or a three would not satisfy the condition. What has gone wrong here is that we have double counted the area of intersection shared by outcomes *E* and *G*. In fact, only four of the six possible outcomes satisfy either *E* or *G*, so the correct answer for the probability of either *E* or *G* occurring is 4/6.

Let us now look at how some of these ideas are expressed using more formal notation. In general, if an event has several possible mutually exclusive outcomes – A, B, C etc – the probabilities are written as follows:

The probability of outcome A resulting is *P(A)*

The probability of outcome B resulting is *P(B)*

and so on.

If we want to find the probability of *either A or B* occurring, we *add* the separate probabilities for *A* and *B*.

$$P(A \text{ or } B) = P(A) + P(B)$$

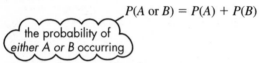
the probability of *either A or B* occurring

(In terms of the circle diagram in Figure 12.4, 'E or G' refers to the combined area covered by the two rings together.)

However, this rule only applies if outcomes A and B are mutually exclusive, otherwise the area of intersection has been double counted.

If two outcomes are not mutually exclusive, the rule must be adjusted to take account of the double counting, as follows.

For events that are not mutually exclusive:

$$P(A \text{ or } B) = P(A) + P(B) - P(A \text{ and } B)$$

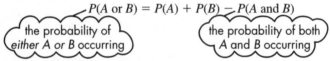
the probability of *either A or B* occurring — the probability of both *A* and *B* occurring

(In terms of the circle diagram in Figure 12.4, 'E and G' refers to the area where the two rings overlap.)

When all the possible outcomes of an event have been listed, the term to describe such a complete list is 'exhaustive'. For example, in Figure 12.4, the combined outcome 'E or G' is not exhaustive because outcomes 1 and 3 are not included. The possible outcomes 'heads' and 'tails' from tossing a coin are mutually exclusive (because if one occurs then the other cannot occur) and they are also exhaustive (together they cover all the possible outcomes).

Independence and multiplying probabilities

Two common terms which often get confused when dealing with probability are 'mutually exclusive' and 'independent'. Two outcomes are independent if the occurrence of one does not affect the likelihood of the other one occurring. What distinguishes 'independence' from 'mutually exclusive' is the sort of situation in which they tend to crop up. Talking about outcomes that are mutually exclusive is most usefully applied where the outcomes concerned arise from a single event. On the other hand, the usefulness of the notion of independence is seen with reference to the outcomes of separate events. For example, the need for introducing ideas of mutual exclusivity in Figure 12.4 arose from dealing with the problem of overlapping outcomes from a single event – the tossing of a die. As will be shown in this section, however, independence is important when we wish to check whether the outcome of one event is likely to affect the possible outcomes of some other event.

A second distinction concerns the sort of calculation that will be involved. The idea of mutually exclusive outcomes tends to be an important consideration where you wish to know about the probability of *either A or B* occurring, and this is a situation where probabilities are being *added*. As you will see shortly, the idea of independence tends to be an important consideration where you wish to know about the probability of *A and B* occurring, and this is a situation where probabilities are being *multiplied*. Here is a simple example to illustrate why we multiply probabilities.

Example
Consider a game where first a coin and then a die are tossed in turn. You win a prize if the outcomes are 'heads' and 'six' respectively. What is the probability of winning?

Solution
The probability of tossing 'heads' with a coin is 1/2 and the probability of throwing a 'six' is 1/6. So the probability of being successful at both events is a half of one sixth or one twelfth. In terms of the original fractions, 1/2 and 1/6, what we have done here is to multiply them. Thus:

$$1/2 \times 1/6 = 1/12$$

So, if there are two separate outcomes, the method of calculating the probability of both occurring is to multiply the individual probabilities. It may seem slightly bizarre to multiply the probabilities when the question seems to be asking about taking two outcomes together (which may imply addition). This distinction between when to add and when to multiply probabilities when they are combined may not be immediately obvious. One simple rule of thumb is to think about whether the process of combining the probabilities has the effect of making the outcome *more* likely or *less* likely. Adding two numbers, even if each number lies between 0 and 1 (as all probability values do), will produce a larger result. Thus, since the chance of *either A or B* occurring is *more* likely than either one on their own, you must add *P(A)* and *P(B)*. However, if you are considering the chance of *A and B* occurring, this is *less* likely than the chance of either *A* or *B* occurring on their own, so you must multiply *P(A)* and *P(B)*. This can be summarized as follows.

Either A or B is ...	... more likely than each one separately ...	... so add the probabilities to get a bigger result
A and B is ...	... less likely than each one separately ...	... so multiply the probabilities to get a smaller result

We now turn to look specifically at the idea of independence, and the next exercise will get you thinking about what independence means.

EXERCISE 12.5. Lottery

Many countries run a national lottery. One way in which this can be organized is that each week, 'numbers' (perhaps available in the range between 001 and 999 inclusive) are sold to the public. Punters might pay something like £1 or £5 each for the number or numbers of their choice. The winning number is drawn from a hat and, after the government has appropriated a suitable profit, the money remaining is split equally among the punters who bought that number.

Suppose you know that last week's winning number was 417, how might that information affect what number you might choose to buy this week?

Comments below

There are two separate elements in this choice. The first is to do with the law of probability (specifically the notion of independence). The key question here is whether the draw for this week is independent of last week's outcome. Provided the draw is done fairly, then the events *should* produce independent outcomes, but you never know ... ! The second aspect to the question relates to the choices you might reasonably expect other people to make. In practice, most people don't believe in the independence of random events and would avoid choosing a number which won the previous week. (This is actually an example of the gambler's fallacy, described earlier.) So, given that all numbers have an equal chance of coming up, 417 would actually be the best choice since there are likely to be fewest people to share the prize with if you bet and won on that number.

EXERCISE 12.6. Tossing and turning

Suppose a fair coin was tossed ten times and each time it showed 'heads'. What would you expect the eleventh toss to show?

Comments below

This question was posed earlier in the chapter, but we can now take advantage of the language of 'independence' by which to explain the answer. Notice that the question states at the outset that the coin is 'fair'. This means that you can assume there is no bias either towards 'heads' or 'tails'. Also, under normal conditions, the successive tosses of a coin are independent events. In other words, the outcomes of the first ten tosses will not influence the outcome of the eleventh toss. Therefore, given that the coin is fair, the chance of heads or tails on the eleventh toss is still 50–50. Of course, if you didn't know that the coin was fair you might wish to argue that there was evidence of a bias towards 'heads', in which case you might feel that 'heads' was more likely on the next throw. However, there is never any argument for thinking that 'tails' is more likely after a run of heads and, interestingly, this is the most commonly held view.

EXERCISE 12.7. Lucky dip

A lucky dip consists of 20 envelopes, only 3 of which contain a prize.

   a) *On the first dip, what is the probability of winning a prize?*
   b) *Suppose the first envelope chosen did contain a prize*

> and the envelope is removed from the bag. What is
> the probability of the second dip also winning a
> prize?
>
> **Comments below**

(a) Using the definition of probability given earlier, the probability of winning on the first dip

$$= \frac{\text{number of prizes}}{\text{number of possibilities}} = \frac{3}{20}$$

(b) After the first dip, the conditions for the second dip have been altered. In the latter case, the required probability

$$= \frac{2}{19}$$

In this instance, the second event (dip number 2) is *dependent* on the first (dip number 1).

A useful way of representing successive events is by using a tree diagram. Here are two examples, the first, Figure 12.5, using a tree diagram to illustrate independent events and the second, Figure 12.6, shows events which are dependent.

As you can see from Figure 12.5, the probability of each outcome is written on the appropriate arrow. In order to find the probability of a particular outcome, say 'HH', simply trace the arrows from left to right and multiply the probabilities on the way.

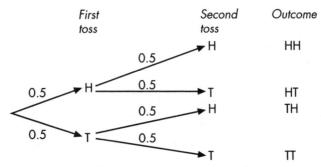

Figure 12.5 A tree diagram showing independent events

EXERCISE 12.8. Using the tree diagram to calculate probabilities with independent events

From Figure 12.5, calculate the probability that if two fair coins are tossed the outcome will be:

    a) *two heads*
    b) *two tails*
    c) *one of each.*

Comments below

  (a) *P(HH)* = 1/2 × 1/2 = 1/4

  (b) *P(TT)* = 1/2 × 1/2 = 1/4

  (c) The probability of getting one of each = *P(HT)* + *P(TH)*
                                      = 1/2 × 1/2 + 1/2 × 1/2
                                        = 1/2

A key feature of this example has been that the probabilities on the arrows corresponding to each second toss are independent of the particular outcomes resulting from the first toss. Figure 12.6 illustrates events which are *not* independent. Note how this time the probabilities on the second set of arrows are *dependent* on the particular outcomes resulting from the first dip.

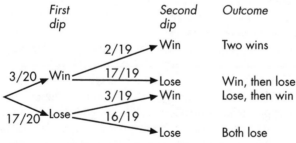

Figure 12.6 A tree diagram showing dependent events

EXERCISE 12.9. Using the tree diagram to calculate probabilities with dependent events

Figure 12.6 represents the 'lucky dip' situation described in Exercise 12.7. Use the tree diagram to calculate the probability that if two envelopes are chosen without replacement, the outcome will be:

a) *two wins*
b) *no wins*
c) *exactly one win.*

Comments below

(a) $P(WW) = 3/20 \times 2/19 = 0.016$

(b) $P(LL) = 17/20 \times 16/19 = 0.716$

(c) The probability of getting exactly one win
$$= P(WL) + P(LW)$$
$$= 3/20 \times 17/19 + 17/20 \times 3/19$$
$$= 0.134 + 0.134 = 0.268$$

The importance of the idea of independence emerges when you are looking at the probabilities involved in a sequence of events – for example, the error factor in a series of runs of a particular machine.

Suppose, for the sake of argument, that the machine in question made identical components and had a chance of 1 in 100 of producing a faulty component on a particular production run. What is the chance of it producing a faulty component on each of three consecutive runs? In theory, you might be tempted to think that the probability was 1/100 each time, so the overall probability is $1/100 \times 1/100 \times 1/100$ or 1 in a million. In practice, if a machine produces a fault, the fault tends to remain unless it is fixed. So, if the machine produces a faulty component on the first run, you might expect that, assuming it is not adjusted between runs, it is progressively *more* likely to behave badly on subsequent runs, giving a calculation of something like $1/100 \times 1/20 \times 1/5$. In other words, with machines, errors tend to cumulate and a series of production runs from the same machine will certainly *not* be independent of each other.

Just as the idea of mutually exclusive outcomes was shown to be important when adding probabilities, independence comes into its own when multiplying probabilities. Again, a few simple examples should help make the point.

EXERCISE 12.10. Three in a row!

    a) *Suppose you toss a fair coin three times in succession. What is the probability of all three coming up 'heads'?*

    b) *Suppose that you were able to take three dips in the lucky dip described in Exercise 12.7. What is the probability that you will win each time?*

    c) *A racing accumulator is the term used to describe a type of bet where you place a different bet on, say, three separate races and win only if all three horses come in first. If the odds on each of your chosen horses are 10–1, what are the odds of winning the accumulator?*

Comments below

    (a) Since successive tosses of a fair coin are independent events, it is perfectly legitimate to multiply each of the separate probabilities involved. Thus:

    Probability of three 'heads' in a row = $(1/2) \times (1/2) \times (1/2) = 1/8$

    (b) As was explained in the comments to Exercise 12.7, the probability of being successful in the second and third dips have altered as a result of the previous outcomes. Thus:

    Probability of winning a prize in all three dips
        = $3/20 \times 2/19 \times 1/18$
        = $6/6840$ (or 1 chance in 1140)

    (c) In this case, it is reasonable to assume that the result of each race is independent of the others. Thus:

    Probability of winning the accumulator = $(1/10) \times (1/10) \times (1/10) = 1/1000$

In summary, then, if two events are independent, the outcome of one does not affect the outcomes of the other.

Summary

Probability is an important topic in its own right. It allows us to understand, calculate and compare the risks around us. This chapter has focused on the basic notion of what probability means and how it is measured, using both *odds* and *statistical probability*. We have examined some common *myths and misconceptions* surrounding probability. The final two sections spelt out the sort of situations where probabilities are combined – situations of the form 'either or' require addition of the separate probabilities, while 'and' requires multiplication. This section also dealt with two important ideas which can easily cause problems when probabilities are added and multiplied – the notion of *mutually exclusive* outcomes and *independence*.

Probability theory also has a valuable role in helping us to draw sensible conclusions from data. In statistics, we attempt to look for patterns in data which show up both similarities and differences. A key role of probability here is to help us decide how likely it is that any observed differences are due, simply, to natural variation. If it is very unlikely that the observed differences can be explained in this way, then we can deduce that the differences are statistically significant. In Chapters 13 and 14 you will see how these ideas of probability and data interpretation are brought together.

Comments on exercises

Exercise 12.1

    (a) £10 × 9/4   = £22.50

    (b) £10 × 16   = £160

    (c) £10 × 5   = £50

    (plus, in each case, your £10 stake back)

Exercise 12.2

EVENT	OUTCOME	NUMBER OF WAYS (n)	NUMBER OF POSSIBILITIES (N)	PROBABILITY $\left(\dfrac{n}{N}\right)$
Tossing a coin	Getting 'tails'	1	2	$\dfrac{1}{2}$
Tossing a die	Getting a score of 5 or more	2	6	$\dfrac{2}{6} = \dfrac{1}{3}$
Choosing a playing card	The suit being 'hearts'	13	52	$\dfrac{13}{52} = \dfrac{1}{4}$
Choosing a playing card	The card being an 'ace'	4	52	$\dfrac{4}{52} = \dfrac{1}{13}$
A baby being born . . .	. . . on a weekend	2	7	$\dfrac{2}{7}$
Someone who was born in December . . .	. . . having their birthday *after* Christmas Day	6	31	$\dfrac{6}{31}$

Exercise 12.4

  (a) Yes *(b)* Yes *(c)* No *(d)* No

13 PROBABILITY MODELS

In the last chapter, you met the basic idea of probability, how it is measured and what sort of meaning we can attach to adding and multiplying probabilities. All this may have seemed very far removed from your general view of what statistics is about. One definition of statistics might be a process of looking for patterns in data and making informed judgements on the basis of quantitative information. So where does probability fit into that process? Unfortunately, many courses in statistics and most textbooks on the subject tend to bundle probability and statistics together with no link made between them. A central aim of this chapter is to explain some of the key ideas of probability – in particular the Normal and binomial distributions – not simply as mathematical curiosities but as essential tools for making decisions in statistics.

Compared to what?

Let us start with a simple medical illustration of how probability supports statistical decision-making. Suppose that you have succumbed to a fever with a high temperature. Your doctor prescribes a drug, which you take and, Hey Presto! the fever has gone next morning. How do you feel about the drug? You may feel sure that it did the trick ('It sure worked for me!'). On the other hand, it is possible that the fever had reached its peak by the time you got round to seeing the doctor, and you would have been well the next morning whether you had taken the drug or not. Even if the drug did work for you, can you say how it might work for other people with a variety of similar symptoms resulting from subtly different causes? What exactly are you comparing it with?

The key question here is the following:

'how likely is it that we would get a result like this from chance alone?'

And this is where probability comes into the decision-making phase of any statistical investigation. But in order to use the approach suggested by the question above, we need to know a bit more about this phrase 'chance alone'. We need to have a good understanding of the range of values that a variable can be expected to take under 'normal' conditions before being in a position to judge whether or not an experimental result is significantly high or low. Depending on the particular circumstances, there are a number of possible 'normal conditions', and they are what are being referred to in the title of this chapter by the phrase 'probability models'. There are a number of useful probability models, including the Poisson and the chi squared distributions, but this chapter looks at just three of the most straightforward, the 'uniform', the 'Normal' and the 'binomial'.

The uniform distribution

Suppose that you decide to test whether a particular die was biased or a random number 'spinner' was truly random. Let us stick with the die and imagine that, after 60 tosses, you got the following observations.

Score	1	2	3	4	5	6
Frequency	5	12	14	6	9	14

Table 13.1   Observed outcomes

On their own, these observed values are just a set of numbers. There needs to be a clear set of 'expected' values against which to compare them. The obvious expectation for an unbiased die is that all outcomes are equally likely, and therefore 60 tosses would yield 10 of each outcome, thus.

Score	1	2	3	4	5	6
Frequency	10	10	10	10	10	10

Table 13.2   Expected outcomes

A bar chart of these expected outcomes will look like Figure 13.1.

Clearly, the pattern here is one of uniformity, hence the name 'uniform distribution'. This uniform distribution, where all the bars are the same height describes a situation like this where there are equally likely outcomes.

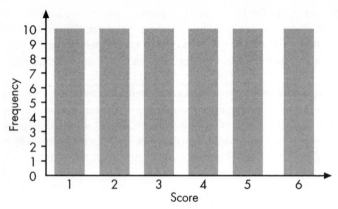

Figure 13.1 Frequency distribution showing the 'expected' outcomes from tossing a die 60 times

---

EXERCISE 13.1. Frequencies versus probabilities

Compare and contrast the uniform frequency distribution in Figure 13.1 with the uniform probability distribution drawn in Figure 13.2.

Comments below

There are two important differences between this probability distribution and the frequency distribution of Figure 13.1. Firstly, the adjacent bars of the probability distribution have been drawn so that they touch. The reason for this is largely one of convenience – an important element in understanding and thinking about probability distributions is to focus on areas of various chunks of graph and this is easier to see when the distribution is continuous rather than discrete. (*Note:* the distinction between discrete and continuous data was explained in Chapter 4.) A second, and more crucial, distinction between the two graphs is that the vertical axis of the probability distribution has been re-scaled so that the uniform height of the rectangle = 0.1667, or 1/6. This leads to an important property of all probability distributions, which you can discover for yourself in the next simple exercise.

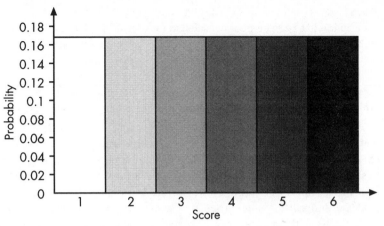

Figure 13.2 Probability distribution showing the 'expected' outcomes from tossing a die

---

EXERCISE 13.2. A fundamental feature of probability distributions

　　a) *Write down the height and the width of this uniform probability distribution in Figure 13.2.*

　　b) *Calculate the area of the combined rectangle.*

Comments below

The point of this exercise is to reveal a key property of all probability distributions, namely that the area bounded by the complete distribution is always equal to 1. They are drawn that way simply for convenience.

A second important property is illustrated in the next exercise.

EXERCISE 13.3. A second fundamental feature of probability distributions

　　a) *If a fair die is tossed once, what is the probability of getting a score in the range 2 to 6 inclusive?*

　　b) *From Figure 13.2, what is the area of the distribution corresponding to a score in the range 2 to 6 inclusive?*

　　c) *What general principle, if any, can you deduce from looking at your answers to parts (a) and (b)?*

Comments follow

The answer to parts (a) and (b) are the same (5/6) because the probability distribution has been cunningly scaled to make it so. In general, the probability that an outcome should lie within the range *a* to *b* corresponds to the area under the probability distribution curve between these two values. For example, the probability of a die score being either 5 or greater corresponds to the combined area of the final two bars of the probability distribution curve in Figure 13.2 (i.e. an area of 2/6 or 1/3).

Finally, let us return to the question which started all this off, namely how we can test whether or not a die is biased. One interpretation of the observed results shown in Table 13.1, albeit a rather silly one, might be to say that, because the observed frequencies are not equal to each other, then the die must be biased. However, this is to apply the comparison with the uniform distribution too rigidly. The rectangular shape of the uniform distribution is simply a *model* of what we might *expect* to happen. We draw it this way because of the equally likely assumptions of the outcomes of a fair die – there is no reason to make any one bar taller than any other if they are all equally likely. However, the reality is that we don't really expect to get this so-called 'expected', perfect outcome! The reason for this apparent paradox is that, due to natural variation (particularly with such a small sample of only 60 throws) we might easily expect one outcome to occur, say, as many as 15 times or perhaps as few as 5 times. But, since we never know in advance which outcome(s) this will be, the best we can offer for our 'model' is the expectation that they will all occur with equal likelihood.

In summary, then, there are two key probability issues that need to be addressed when decisions are to be made about whether or not a batch of observed data indicate a significant pattern. These are as follows.

- Compared to what? You need to decide what are the 'expected' values against which the observed values are to be compared. In the case of things like testing dice or random spinners for bias, the 'expected' behaviour is one of equal likelihood and so the model probability distribution will be 'uniform'.

- Natural variation. Even if the die or the spinner were truly random, you wouldn't expect a particular sample of observations to produce an exact replica of the underlying uniform model. Due to natural variation there will inevitably be fluctuations. The crunch question when examining a set of observed results becomes 'how likely is it that I would get a fluctuation as wide as this from chance alone?' and this is followed up in the final chapter 'Testing for a difference'.

The Normal distribution

You may have felt that the uniform probability distribution described in the last section was relevant to artificial situations like tossing dice, etc, but wasn't a very useful model of reality. This is a fair criticism, but it is a simple example of a distribution and provides a useful introduction to probability models in general. However, in the 'natural' world where measurement involves finding the sizes of plants, animals and humans, a rather different sort of distribution tends to predominate.

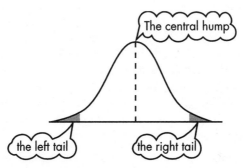

Figure 13.3 The Normal curve

The 'bell-shaped' curve in Figure 13.3 is one that accords with the probability distribution known as the Normal curve. The characteristic shape of the Normal curve is that it has a hump in the middle and two tails, one on either side (easily distinguishable, of course, from the Bactrian or two-humped camel, which is the exact opposite!). Notice that the term 'Normal' has been written with a capital 'N'. This is in order to stress that a specialized use of the term 'Normal' is being employed here. It is worth noting that the Normal curve is not just any old bell-shaped curve with a central hump and a tail on either side. In fact it has a very specific curvature and is defined by an exact mathematical formula.

Other names for this distributional curve include:

- the Gaussian distribution or the Laplace–Gauss law (after the mathematicians, Laplace and Gauss, who investigated and defined it),
- the law of error curve,
- the bell-shaped curve,
- the Z curve.

It is quite remarkable how many natural phenomena appear to be based on underlying distributions with this general shape. Here are just a few examples:

- the heights of a sample of adults,
- the weights of a harvest of plums from a tree,
- the errors in a batch of engine crankshafts,
- the daily number of road accidents in the UK measured over a year,
- the expected life of a batch of light bulbs,
- the germination time of the seeds in a packet,
- the scores of students on tests,
- the lengths jumped by a long jumper, taking all her jumps over one season,
- the weights of a batch of eggs collected over a period from a farm,
- the Intelligence Quotient (IQ) of a random sample of people as measured by the same test.

Let us now look at just one of these variables in more detail – the Intelligence Quotient (IQ) of a random sample of people. Whether or not you believe that IQ tests are valid measures of people's intelligence, if you were to test, say 100 people randomly selected from the population, using the same 'standardized' IQ test, it is likely that the distribution of the results would be roughly Normal with a mean close to 100 and a standard deviation close to 15. The reason for this is simple. IQ tests are *designed* to produce such a distribution. In fact, the reason that they are called 'standardized' tests is that they have been adjusted so that they produce this sort of result. Of course, if only a small random sample is chosen, then the picture is unlikely to be obvious. Table 13.3 shows a typical set of scores for a sample of ten randomly chosen people.

IQ RANGE	NUMBER OF PEOPLE
IQ < 70	0
70 ≤ IQ < 85	2
85 ≤ IQ < 100	3
100 ≤ IQ < 115	2
115 ≤ IQ < 130	3
130 ≤	0
TOTAL	10

Table 13.3   Distribution of the IQs of ten people chosen at random

Plotting these as a histogram gives the following picture.

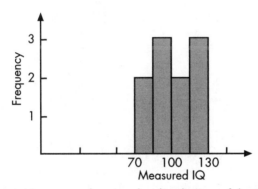

Figure 13.4   Histogram showing the distribution of the measured
IQs of ten people

There isn't much of a clear picture here, in fact. You may have been hoping for a nice symmetrical bell-shaped curve, but what we have is more of a camel. The reason for this is that, although IQs are basically Normal in their distribution, a sample size of ten is too small to expect this underlying model to show up clearly. Let us now increase the sample size to 100 and see what happens to the overall pattern, see Table 13.4.

The histogram of this distribution is shown in Figure 13.5.

Clearly this is beginning to come closer to the bell-shaped curve that was predicted. You might imagine that if the sample size were increased substantially, say to 1000, the distribution would become even more symmetrical and bell-shaped. It might look something like Figure 13.6.

IQ RANGE	NUMBER OF PEOPLE
IQ < 70	5
70 ≤ IQ < 85	10
85 ≤ IQ < 100	38
100 ≤ IQ < 115	30
115 ≤ IQ < 130	13
130 ≤ IQ	1
TOTAL	100

Table 13.4   Distribution of the IQs of 100 people chosen at random

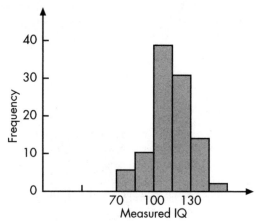

Figure 13.5   Histogram showing the distribution of the measured IQs of 100 people

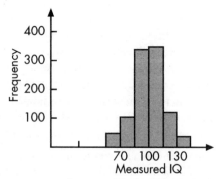

Figure 13.6   Histogram showing the distribution of the measured IQs of 1000 people

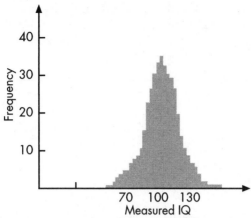

Figure 13.7   Histogram showing the distribution of measured IQs of 1000 people, grouped into narrow intervals of three IQ units

However, there is still a problem with the fact that the Normal curve is a smooth continuous function whereas these histograms are all stepped like stairs. No matter how large a sample size we take, the histogram will still lack the smooth appearance of the Normal curve. This problem can be reduced by grouping the data into much narrower intervals. In this example, we might wish to reduce the width of the intervals from 15 IQ units to, say, 3, producing a histogram like Figure 13.7.

In the limiting case, where the sample is infinitely large and the class intervals are infinitesimally small, the Normal curve and the histogram coincide. So, once again the idea of a 'model' is appropriate as a way of thinking about probability distributions, in that the Normal curve is the underlying model for these histograms.

An important property of the Normal distribution is that it can be defined entirely from knowing just two key facts – where it is centred (based on the *mean*) and how widely spread it is (as measured by the *standard deviation* and the square of the standard deviation, the *variance*). Details of how to calculate the mean the variance and the standard deviation were given in Chapter 5.

The shorthand way of describing this distribution of IQs is simply to state the mean and the variance, thus:

$$N(100,225)$$

This is a neat way of saying that the distribution is Normal (the 'N' part), with a mean value of 100 and a variance of 225 (i.e. a standard deviation of $\sqrt{225} = 15$). In general, a Normal distribution would usually be stated in this format, as follows:

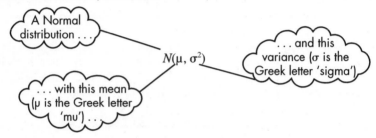

Calculating probabilities from the Normal curve

We are now going to put the Normal curve to work by using it to calculate probabilities. The next exercise will give you the basic idea of what this involves.

EXERCISE 13.4. Interpreting the Normal curve

Use Figure 13.7 to make the following estimates.

   a) *Roughly what proportion of the population would have a measured IQ of less than 100?*

   b) *The 95% interval refers to the range of values within which the middle 95% of the distribution lies. What range of IQs, very roughly, define the 95% interval for this distribution?*

   c) *Someone with a measured IQ in excess of 140 is considered to be eligible for 'MENSA'. What is the probability that a randomly chosen person falls into this category?*

Comments below

You may have noticed two things about this distribution. Firstly, the mean is 100 and secondly that it is more or less symmetrical. This suggests that roughly half of the IQ values will be on either side of 100. Next, finding the middle 95% of a distribution involves cutting off two tails of $2\frac{1}{2}\%$ on either side. This might suggest something like the range

70 to 130. Finally, the cut-off value for defining MENSA levels produces a very small tail indeed – so small that it is not really possible to estimate here, but something like 1% might be a reasonable guess.

Provided that you know the values of both the mean and standard deviation of a Normal distribution, you are able, with the aid of Standardized Normal Tables, to make these sorts of predictions much more accurately. As you can see from the wording of the different parts of the previous exercise, this may involve thinking about sections of the curve in two rather different ways; firstly in terms of the *proportions of the population* falling within a particular range of values (part (a) above) or the *probability* of a randomly chosen item falling within a particular range (part (c) above). Either way, the calculation is essentially the same and involves using Standardized Normal Tables to match up ranges of values of the variable in question with corresponding areas of the curve.

z	$F(z)$	z	$F(z)$
-3.0	0.001	0.0	0.500
-2.8	0.003	0.2	0.579
-2.6	0.005	0.4	0.655
-2.4	0.008	0.6	0.726
-2.2	0.016	0.8	0.789
-2.0	0.023	1.0	0.841
-1.8	0.036	1.2	0.885
-1.6	0.055	1.4	0.919
-1.4	0.081	1.6	0.945
-1.2	0.115	1.8	0.964
-1.0	0.159	2.0	0.977
-0.8	0.211	2.2	0.986
-0.6	0.274	2.4	0.992
-0.4	0.345	2.6	0.995
-0.2	0.421	2.8	0.997
		3.0	0.999

Table 13.5 Simplified version of the Normal distribution function table

Table 13.5 shows a simplified form of the Normal distribution function (often described as the Z function). A key explaining what the numbers refer to is given on page 240. Note that this table allows you to look up z values accurate only to one decimal place. For a more complete table, you will need to consult the appendix of an advanced statistics textbook.

Key

For a particular value, X:

z = the number of standard deviation units that X lies to the right of the mean

$F(z)$ = the area under the curve to the left of X.

The Normal distribution

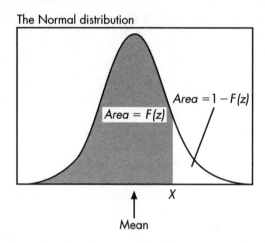

Area = $1 - F(z)$

Area = $F(z)$

X

Mean

Using the Standardized Normal Tables – a worked example

Question

The measured IQs of the population are Normally distributed according to $N(100,225)$. Use the Standardized Normal Tables to calculate:

    (a)   the proportion of people whose measured IQ is less than 115;

    (b)   the proportion of people whose measured IQ is under 85;

    (c)   the 95% interval;

    (d)   the probability that a randomly chosen person falls into the category 'eligible for MENSA' (i.e. measured IQ > 140).

Solution

Notice that these questions are similar to those posed in Exercise 13.4, except that they are phrased more in the 'standardized Normal jargon' of a statistics textbook. Note also that you are expected to find the areas under the curve with reference to the tables, rather than simply by eye. The first crucial number we need is the standard deviation. $SD = \sqrt{225} = 15$.

- (a) The IQ value 115 is 1 standard deviation unit to the right of the mean, so $z = 1$. From Table 13.5, the area to the left of this value is 0.841, or about 84%.

- (b) The IQ value 85 is $-15/15$, or 1 standard deviation unit to the left of the mean, so $z = -1.0$. From the table, this corresponds to an area of 0.159, or about 16%.

- (c) Finding the middle 95% interval is a bit trickier. First, you need to observe that this range corresponds to two $F(z)$ values at $2\frac{1}{2}\%$ and $97\frac{1}{2}\%$ respectively. Focusing just on the $97\frac{1}{2}\%$ (or 0.975) we can find its corresponding z value by looking up 0.975 backwards in the table. The closest we can get is a z value of 2 (for $z = 2$, $F(z) = 0.977$), and this gives the following extremely useful general result for all Normal distributions:

    > Roughly 95% of the values lie within 2 standard deviations on either side of the mean.

    In this example, 2 standard deviations on either side of the mean gives a range of 100 ± 30, or between 70 and 130.

- (d) A measured IQ of 140 lies $2\frac{2}{3}$, or roughly 2.6, standard deviation units to the right of the mean. From the table, the area to the left of $z = 2.6$ is 0.995. However, we are interested in the proportion of the population with measured IQs above 140, so we need the area of the little tail to the right. In other words, required area $= 1 - 0.995 = 0.005$. This suggests that roughly 5 people in 1000 are likely to score sufficiently highly in a standardized test to be accepted as a member of MENSA.

You will need some practice at using Normal tables, so do Exercise 13.5 now.

EXERCISE 13.5. Calculating with the Normal tables

a) *Babies are classed as being 'low birth weight' if they weigh less than 2500 g at birth. Assuming that the birth weights of babies are Normally distributed with a mean of 3500 g and standard deviation 400 g, roughly what proportion of babies would be expected to fall into the low birth weight category?*

b) *The expected weight of crisps of a certain brand of potato crisps, in grams is N(28,9). What should the 'guaranteed minimum weight' be set at so that only 1% of the bags could be expected to fall under this weight?*

c) *A particular police force rejects 20% of its applicants because they are too short in height. If the heights of the applicants, in cm, are N(170,16), can you deduce what minimum height the police force is applying?*

Comments on page 252

Finally, to end this section, here is a summary of the key features of the Normal distribution.

   (a)   the area under the curve $= 1$

   (b)   the probability that X lies between a and b, $p(a < X < b)$, is represented by the area under the curve bounded by a and b.

   (c)   Mean $= \mu$, Standard deviation $= \sigma$

The Normal distribution

$p(a < X < b)$

a b

The binomial distribution

The binomial distribution is the third and final probability model described here. Unlike the Normal distribution which is used to model continuous data like height, weight, time, and so on, the binomial distribution models events which have two discrete outcomes. Typically it describes situations where there are repeated trials, each of which can either turn out to be a 'success' or a 'failure'. Here are a few examples.

EVENT	TWO POSSIBLE OUTCOMES
The composition of boy/girl in a family of, say three children	Each child can be either a 'girl' or a 'boy'
The outcomes, on several tosses of a coin	Each outcome can be either 'heads' or 'tails'
The colour of flowers (either red or yellow) grown from a packet of flower seeds	Each seed can turn out either 'red' or 'yellow'

The terms 'success' and 'failure' may not seem highly appropriate to all of these situations. After all who is a statistician to pronounce on whether the sex of a baby, whether a girl or a boy, should be a 'success' or a 'failure'? However, the terms are not used here as value judgements about the social merits of the outcomes. The use of the terms is merely a useful device for stressing that there are only two possible outcomes. In other words, for a particular binomial event:

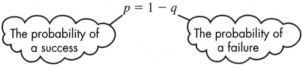

$$p = 1 - q$$

The probability of a success

The probability of a failure

To express these ideas in the language of the previous chapter, in the binomial distribution we are dealing with a set of trials, each of which has only two possible mutually exclusive and exhaustive results, 'success' and 'failure', with probabilities p and $1 - p \; (= q)$.

We shall take as our working example the question of family composition.

EXERCISE 13.6. Sisters and brothers

Try to think of eight families that you know which have three children. Make a table like Table 13.6, summarizing the girl/boy composition of each family.

FAMILY NAME	SEQUENCE	GIRLS	BOYS
Graham	B G G	2	1
Lunn	G G B	2	1
Ware	B B B	0	3
Baker	G G G	3	0
Staunton	G G B	2	1
Cheriyan	B B G	1	2
Liang	B B G	1	2
Ostrowski	G B G	2	1

Table 13.6   Girl/boy composition of eight families

Now make a tally of the number of each separate outcome, and sketch the corresponding bar chart, as in Table 13.7 and Figure 13.8.

OUTCOME G	B	TALLY	FREQUENCY
0	3	I	1
1	2	II	2
2	1	IIII	4
3	0	I	1

Table 13.7   Tallying the girl/boy combinations

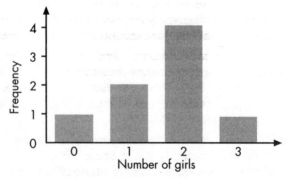

Figure 13.8   Charting the tallies

Although your bar chart won't necessarily look identical to the one in Figure 13.8, it is likely that it will have the same familiar pattern. This is the tendency to show a peak in the middle and to tail off at either side – a characteristic also of the Normal curve. Unlike the Normal curve, however, the bars are separated, emphasizing that the binomial distribution is a way of modelling *discrete* outcomes.

Clearly, this central peaking of the distribution indicates that middle values (in this case, families with 1 or 2 girls) are more likely than extreme cases (families with 0 or 3 girls). But why should this be? The clue to answering this question is revealed by examining some particular families in more detail. Let us take the four families with 2 girls and 1 boy (Graham, Lunn, Staunton and Ostrowski). There are three separate ways in which this outcome (two girls and one boy) can be produced, all of which are represented in these families. They are:

BGG, GBG and GGB.

Similarly, there are three separate ways in which families with two boys and one girl can be produced. They are:

GBB, BGB, and BBG.

Contrast these with all-boy and all-girl families, each of which can be produced in only one way – BBB and GGG respectively.

These outcomes can be summarized in a tree diagram, as shown in Figure 13.9.

In this figure, the probability of each birth outcome is written beside the appropriate line and the overall probability for a particular combination is given on the right of the diagram. These results represent the 'expected' outcomes and are given the label 'binomial distribution', see Table 13.8.

The corresponding binomial probability distribution for three-children families is identical to this, but the frequencies have been scaled down so that they add to 1, see Figure 13.9. (Remember from earlier in this chapter a key requirement of all probability distributions is that their area must equal 1.)

Just as there is a shorthand way of describing any Normal distribution in terms of two numbers (the mean and variance), so can we define any binomial distribution. Here the two key pieces of information are 'n', the number of trials and 'p', the probability of 'success'.

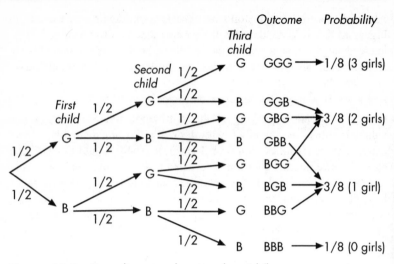

Figure 13.9 Tree diagram showing the girl/boy outcomes in a three-children family

| OUTCOME | | EXPECTED |
G	B	FREQUENCY
0	3	1
1	2	3
2	1	3
3	0	1

Table 13.8   Theoretical outcomes for eight three-children families – the binomial distribution

| OUTCOME | | PROBABILITY |
G	B	
0	3	1/8
1	2	3/8
2	1	3/8
3	0	1/8

Table 13.9   Binomial probability distribution for three-children families

Figure 13.9 showed how the binomial distribution could be demonstrated diagrammatically, with the aid of a tree diagram. To end this section on the binomial distribution, here is a brief explanation of how these ideas link to an important idea in mathematics known as the binomial expansion. The expansion itself is first described and then there follows an explanation of how it illuminates the binomial distribution.

The binomial expansion

A binomial expression is an expression of the form $(x + y)^n$. For example, here are some binomial expressions.

$$(x + y)^2$$
$$(x + y)^3$$
$$(x + y)^4 \text{ and so on.}$$

Of course, the letters need not be x and y. The following are also binomial expressions:

$$(p + q)^5$$
$$(a + b)^9$$

So, there are two main characteristics of a basic binomial expression, namely:

■ there are *two* letters (that explains the 'bi-' in the word 'binomial') inside the brackets with a '+' between them (this can also be a '−').

■ the whole bracket is *raised to a power* – perhaps squared, cubed, etc. This is the 'n' in the binomial expression.

The essence of the binomial expansion is that it provides a simple rule to enable you to expand out the brackets quickly, should you wish to do so. We will now look at some of these expansions in order to investigate what the rule might be, starting with the simple case where $n = 2$.

Expanding out the expression $(x + y)^2$ gives the following result:

$$(x + y)^2 = (x + y)(x + y) = x(x + y) + y(x + y)$$
$$= x^2 + xy + xy + y^2$$
$$= x^2 + 2xy + y^2$$

Notice the patterns in this particular expansion. Firstly there are three terms, an x^2, an xy and a y^2 term. Secondly, the middle term (i.e. the xy term) has occurred twice, while the first and third terms occurred only once. So, the pattern of the coefficients, in sequence, is 1, 2, 1. We are now going to raise the power of the binomial expression from two to three and see what effect this has on these patterns of the expansion.

You might like to check for yourself that the expansion of $(x + y)^3$ gives the following result:

$$(x + y)^3 = x^3 + 3x^2y + 3xy^2 + y^3$$

This time the pattern of the terms of the expansion reveal four term-types, x^3, x^2y, xy^2 and y^3. The corresponding coefficients run 1, 3, 3 and 1.

Increasing n to 4, 5 etc and expanding out all the brackets is clearly going to be a bit of a slog but, given time, it could be done. Table 13.10 gives the term-types and corresponding coefficients for the binomial expansions of $(x + y)^n$ for values of n from 1 up to 5.

n	TERM-TYPES	COEFFICIENTS
1	x, y	1, 1
2	x^2, xy, y^2	1, 2, 1
3	x^3, x^2y, xy^3, y^3	1, 3, 3, 1
4	$x^4, x^3y, x^2y^2, xy^3, y^4$	1, 4, 6, 4, 1
5	$x^5, x^4y, x^3y^2, x^2y^3, xy^4, y^5$	1, 5, 10, 10, 5, 1

Table 13.10 Term-types and the corresponding coefficients of the binomial expansion of $(x + y)^n$ for values of n from 1 up to 5

From this table, it is a simple task to write down the binomial expansion of, say $(x + y)^5$. It involves attaching each term-type to its corresponding coefficient, thus.

$$(x + y)^5 = x^5 + 5x^4y + 10x^3y^2 + 10x^2y^3 + 5xy^4 + y^5$$

For binomial expansions involving values of n larger than 5, you would need to expand this table.[1] For example, if you wish to expand $(x + y)^6$, the first step is to identify the term-types. Notice the pattern of term-types that is evident for smaller values of n – as you can see, for each successive term, the power of x reduces by one while the power of y increases by one. Maintaining this pattern for $n = 6$, this would suggest the following term-types:

$$x^6, x^5y, x^4y^2, x^3y^3, x^2y^4, xy^5, y^6$$

[1] There is a more sophisticated method of generating these coefficients which involves calculating 'factorials', but it is beyond the scope of this book.

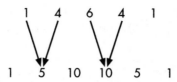

Figure 13.10 Using Pascal's triangle to generate binomial coefficients

To work out the corresponding coefficients, notice that the coefficients in Table 13.10 are arranged in a triangle. This arrangement is known as Pascal's triangle. Here, too, there is an interesting pattern in the way these coefficients are generated.

Notice how each number in a particular row in Table 13.10 is generated by the two numbers immediately above it. So, for example, the second number (i.e. the 5) in the sequence for $n = 5$ above can be generated by adding the 1 and the 4 immediately above it. This is illustrated in Figure 13.10. Similarly, the 10 in the same row in the diagram can be generated by adding the 6 and the 4 immediately above it, and so on for all the numbers in the triangle.

EXERCISE 13.7. Generating the next binomial expansion
  a) *Using the method illustrated in Figure 13.10, calculate the coefficients for* n = 6.
  b) *Now write out the binomial expansion for* $(x + y)^6$.

Comments on page 253

Linking the binomial distribution and the binomial expansion

You may be wondering what the binomial expansion – an apparently abstract piece of algebraic manipulation – has to do with the sort of examples which were introduced earlier as illustrative of the binomial distribution. The link should become apparent if we return to the original example of the theoretical distribution of girls and boys in a family of three children. Let us represent the probability of a child being born a girl, with the letter 'g' and the corresponding probability of a child being born a boy, with the letter 'b'. The situation can be expressed with a binomial expression, as follows.

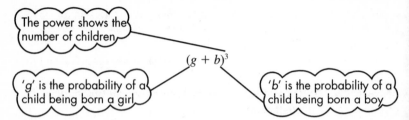

The real pay-off comes when the binomial expression is expanded, as follows:

$$(g + b)^3 = g^3 + 3g^2b + 3gb^2 + b^3$$

Now, each term in the expansion corresponds to the various girl/boy combinations in a three-children family. Not only that, but if we were to substitute the probability values for 'g' and 'b' into the expansion, the term values provide the theoretical probabilities of each separate outcome. Since the probabilities of 'g' and 'b' are, in practice roughly (not exactly) equal, we will take $g = b = 1/2$. Table 13.11 shows each term in more detail and explains what information it provides in this particular context.

As was indicated at the beginning of this section on the binomial distrib-

TERM	MEANING	VALUE WHEN $g = b = 1/2$
g^3	This term arose by multiplying $g \times g \times g$ and therefore corresponds to a family of three girls.	$(1/2)^3 = 1/8$
$3g^2b$	This term arose by multiplying $g \times g \times b$ and therefore corresponds to a family of two girls and one boy. Notice that there are three ways that such a family could be produced – GGB, GBG and BGG – hence the coefficient of 3.	$3 \times (1/2)^3 = 3/8$
$3gb^2$	This term arose by multiplying $g \times b \times b$ and therefore corresponds to a family of one girl and two boys. Again, the three possible ways – BBG, BGB and GBB – produce the coefficient of 3.	$3 \times (1/2)^3 = 3/8$
b^3	This term arose by multiplying $b \times b \times b$ and therefore corresponds to a family of three boys.	$(1/2)^3 = 1/8$

Table 13.11

ution, most textbooks use the letters 'p' and 'q' to describe the probabilities of 'success' and 'failure', respectively and the letter 'n' represents the number of 'trials'. It is always useful to know the mean and variance of any distribution. The mean of the binomial distribution = np, and the variance = npq. (You will need to consult a more advanced statistics book if you wish to know how these results are derived.) So in this example, for a family of 3 children:

The mean number of girls = np = 3 × 0.5 = 1.5.

The variance of the number of girls = npq = 3 × 0.5 × 0.5 = 0.75.

Finally, here is a summary of the key features of the binomial distribution.

    (a)   the area under the histogram = 1

    (b)   the probability that X lies between a and b, $p(a < X < b)$, is represented by the area under the histogram bounded by a and b

    (c)   Where p = probability of success and q = probability of failure, $p + q = 1$

    (d)   The probabilities for each outcome can be calculated by expanding the binomial expression $(p + q)^n$

    (e)   Mean = np, Variance = npq.

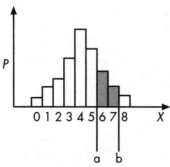

Based on the expansion of
$(p + q)^8$

Summary

This chapter has attempted to create a bridge between probability and statistics by describing, briefly, some of the best known probability models – specifically the 'uniform', the 'Normal' and the 'binomial' probability distributions. As will be demonstrated in the final chapter of the book, knowing about these models and how they vary is the basis on which we are able to make judgements about patterns in data collected as part of an investigation.

Certain key features are common to all probability models, as follows.

■ They are theoretical or 'perfect' representations and we never expect a particular sample to have this exact distribution, especially if the batch size is small.

■ By convention, the scale is adjusted so that the area under the distribution $= 1$.

■ The probability that any randomly chosen value, X, lies between a and b, $p(a < X < b)$, is represented by the area under the curve bounded by a and b.

Comments on exercises

Exercise 13.5

(a) Standard deviation $= 400$ g. A birth weight of 2500 falls 1000 g to the left of the mean. So, $z = -1000/400 = -2.5$. From the Normal tables, this corresponds to an $F(z)$ value of around 0.006 (roughly).

This suggests a 'tail' of about 0.006, i.e. roughly six babies in every thousand are expected to be 'low birth weight'.

(b) Standard deviation $= \sqrt{9} = 3$. $F(z) = 0.01$. From the Normal tables, this corresponds to a z value of around -2.4. When $SD = 3$, this places the value of X at $-2.4 \times 3 = 7.2$ g to the left of the mean.

So, the guaranteed minimum weight $= 28 - 7.2 = 20.8$ g.

(c) Standard deviation = $\sqrt{16} = 4$. $F(z) = 0.2$. From the Normal tables, this corresponds to a z value of around -0.85. When $SD = 4$, this places the value of X at $-0.85 \times 4 = 3.4$ cm to the left of the mean.

So, minimum height applied by the police force = 170 cm $-$ 3.4 cm = 166.6 cm.

Exercise 13.7

(a) The coefficients are 1, 6, 15, 20, 15, 6, 1.

(b) $(x + y)^6 = x^6 + 6x^5y + 15x^4y^2 + 20x^3y^3 + 15x^2y^4 + 6xy^5 + y^6$

14 DECIDING ON DIFFERENCES

> *'Is there any point to which you would wish to draw my attention?'*
> *'To the curious incident of the dog in the night-time.'*
> *'The dog did nothing in the night-time.'*
> *'That was the curious incident,'* remarked Sherlock Holmes.
>
> **From *Silver Blaze*, by Sir Arthur Conan Doyle**

In the past, post-Christmas news broadcasts have tended to include the number of people who had died each day over the Christmas period on the roads. This was a practice which was followed for a number of years until it was pointed out that, rather than the figures being exceptionally high, they tended to be considerably *lower* than typical daily figures for road fatalities over the rest of the year. Like Sherlock Holmes' silent dog, events or statistical data are only 'significant' if they contrast with the typical pattern. To be told in a newspaper that 'the total number of fatal workplace accidents on Wednesdays was 262' in a particular year gives no clue as to whether this is a lot or a little compared to the other days in the week. However, when you discover that the corresponding figure for Fridays was 99, there are immediately some questions to be asked about whether this degree of variation is greater than would be expected by chance alone and, if so, why, and how can the Wednesday figure be reduced?

As the title suggests, the central theme of this chapter is about how we can decide whether or not observed differences can be described as being 'statistically significant' (in other words, indicate some real underlying difference) or simply the result of chance variation. There are a variety of formal methods designed to test differences and two of them – the 'z test' and the 'binomial test' – will be outlined here. These two tests relate closely to the ideas introduced in the previous chapter on the Normal distribution and the binomial distribution, respectively, and you should read Chapter 13 before embarking on this chapter.

As you might anticipate, testing for differences is a major component of statistical work and there is only space here for a brief introduction to some of the main ideas. For a fuller treatment, you will need to consult a more advanced textbook and read the section entitled either 'tests of significance' or 'hypothesis tests'.

Picturing differences

Before applying formal statistical methods as a means of analysing data (for example, calculating means and standard deviations or correlation coefficients etc), it is usually a good idea to plot the figures on some sort of graph. This can be helpful in a number of respects:

- getting a quick visual impression of the overall distribution;
- spotting extreme values that may suggest faulty data; or
- revealing interesting patterns that may be worth investigating further.

The following example illustrates the point that a picture is sometimes worth a thousand words. It is also presented here to raise the question of what factors need to be considered when deciding on differences.

EXERCISE 14.1. Whose egg?

Duck and hen eggs can sometimes look fairly similar, although duck eggs are usually heavier.

If you are given a suspicious egg weighing 72 g and access to a batch of 'typical' hen egg weights, how might you set about deciding whether or not the 'suspicious egg' really was laid by a hen or a duck?

Comments below

Let us assume that you are able to weigh a batch of, say, 100 genuine hen eggs. The 5-figure summary below indicates the general features of this batch.

Figure 14.1   5-figure summary of a batch of 100 hen egg weights (g)

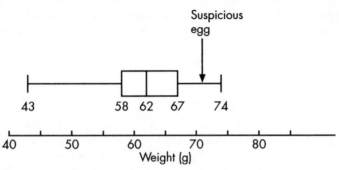

Figure 14.1   Boxplot of the data in Figure 14.1

Figure 14.2 shows these data drawn in a boxplot. (Boxplots were explained in Chapter 5.)

Representing the data in this way, you do get a much clearer sense of the overall shape of the distribution than from the raw data values alone. Furthermore, you also get a good intuition about how likely it is that this suspicious egg might have been laid by a hen. Certainly there are one or two extremely heavy hen eggs which weigh 72 g or more, but such a weight is very unusual for a hen egg and you might be inclined to think that this is more likely to be a duck egg. However, before rushing to this conclusion, you would need to have a rough idea of the distribution of duck egg weights. If the typical pattern of duck egg weights looks something like that shown in Figure 14.3, then, indeed, a duck egg looks more

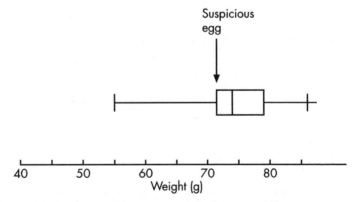

Figure 14.3   A possible pattern of duck egg weights

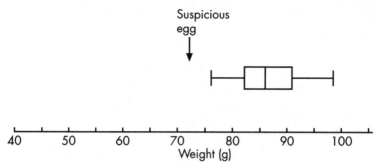

Figure 14.4 An alternative pattern of duck egg weights

likely. If, however, the pattern of duck egg weights is more like that shown in Figure 14.4, then it is less likely to be duck than hen.

Let us now see what the decision would have been if all the hen eggs had been, say, 5 g heavier than indicated in Figure 14.1.

EXERCISE 14.2. Whose egg 2?

The 5-figure summary below indicates the general features of a heavier batch of hen eggs than before.

n = 100

	67	
63		72
48		79

Figure 14.5 5-figure summary of a heavier batch of 100 hen egg weights (g)
Source: fictitious

Draw the boxplot corresponding to the data in Figure 14.5 and try to decide whether or not there is now evidence that the suspicious egg was not laid by a hen.

Comments follow

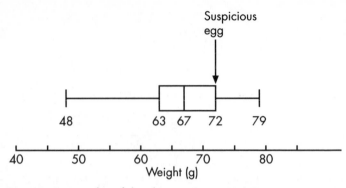

Figure 14.6 Boxplot of the data in Figure 14.5

The boxplot corresponding to the data in Figure 14.5, is shown in Figure 14.6.

The effect of adding 5 g to the weights of each egg has been to move the boxplot five units to the right. The suspicious egg now no longer seems out of place in this distribution. In fact, the weight of the suspicious egg now corresponds with the value of the upper quartile, which suggests that around 25% of hen eggs are likely to be this weight or more. We might conclude, then, that there is not sufficient evidence to believe that this is not a hen egg. Note that this last sentence, which has been written with a double-negative, may sound an unnecessarily complicated way of expressing a simple idea. However, it does mean something different to what you get if you drop both the 'nots' in the sentence. In other words, we have not proved that it is a hen's egg, but rather we have been unable to prove that it is not!

Finally, let us consider a third possibility, namely that the distribution of hen egg weights was centred around a median of 67 g (as for the previous example) but this time with a greatly reduced spread.

EXERCISE 14.3. Whose egg 3?

The 5-figure summary below indicates the general features of a batch of hen eggs similar in average weight to the last one, but with a much narrower spread.

	67	
65		70
n = 100		
58		73

Figure 14.7   5-figure summary of a batch of 100 hen egg
weights (g) – heavy eggs, narrow spread
Source: fictitious

*Draw the boxplot corresponding to the data in Figure 14.7 and
try to decide whether or not there is now evidence that the suspicious egg was not laid by a hen.*

Comments below

The boxplot depicting the data in Figure 14.7, is shown in Figure 14.8.

This third situation is particularly interesting. This time the suspicious
egg falls in the upper whisker of the boxplot and close to the upper
extreme value of 73 g. So, in short, it barely falls into the distribution of
egg weights at all. Despite the fact, then, that the suspicious egg is now
being compared with heavier hen eggs, we are this time inclined to reject
the theory that the unknown egg was laid by a hen. The key factor here is
that the narrow spread of hen egg weights provides a much tighter interval of weight as a criterion for deciding whether or not an egg was laid
by a hen.

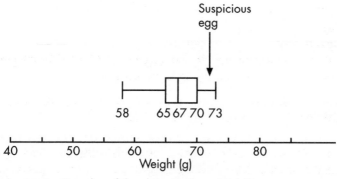

Figure 14.8   Boxplot of the data in Figure 14.7

The point of this simple example has been to illustrate the two key elements that need to be considered when comparing a particular value with a larger batch of values. These are:

- ■ what is a typical or average value of the batch?
- ■ how widely spread is the batch?

It is on these two considerations that much decision-making in statistics is based.

Contexts for investigating differences

A quick glance through any daily newspaper will produce a variety of situations where the process of investigating differences has occurred. Table 14.1 shows a few examples.

The common theme in these three health surveys, and in many other newspaper reports like them, is that researchers have made some sort of comparison between two or more things and 'proved' that there is a 'significant difference' between them – i.e. a difference that could not merely be explained by natural variation in the data. The notion of what constitutes 'proof' is worth examining more closely here. A statistician's

HEADLINE	BASIC STORY
Top doctors condemn alternative medicine	Members of the Royal College of Physicians claim that 'pseudo-scientific' alternative medicine is often an irrational waste of time and money and can even be a serious risk to health. 'People have been misled by the placebo effect, which comes from suggestion, not medical intervention', they claimed.
Child smokers are more likely to take drugs, survey finds	According to a Health Education Authority survey, more than half the children who smoke regularly have been offered drugs and half have tried them. They are also more likely to drink alcohol. The survey, conducted by MORI, was based on interviews with 10 000 nine to 15-year olds.
Most heart disease in North	According to a Health Education Authority report, a survey into coronary heart disease has confirmed that death rates are highest in the Northern, Yorkshire and North-western regions and lowest in the Thames, Wessex, Oxford and East Anglia regions.

Table 14.1 Examples of 'differences' that crop up in newspapers

view of 'proof' is rather different to that of the mathematician. Whereas a mathematical result is normally 'proved' with 100% certainty, a statistical proof is closer to the legal notion of a conclusion being proved 'beyond reasonable doubt'. The difference is that mathematicians are able to create inside their heads a perfect world of numbers, symbols and relationships in which absolute truth can exist. Statisticians, like juries, have to take the real world as they find it and make the best decisions they can under conditions of uncertainty. All statistical judgements, therefore, are given in association with a measure of probability attached to them, which indicates how confident we might be that the pattern is a real one and not just some fluke result due to chance. However, we can never rule out the possibility that it just might be a fluke result.

Some basic terminology

This section provides an introduction to the key ideas and vocabulary that you need to know when deciding on differences in statistics. Let us now use the egg example as a context for defining some of the key words in this area of statistics. We shall assume that the egg weights are Normally distributed (see Chapter 13 for an explanation of the Normal distribution) with a mean of 62 g and a standard deviation of 4 g (mean and standard deviation are explained in Chapter 5). We are investigating whether or not a 72 g egg is likely to have come from this distribution. Figure 14.9 sets the scene.

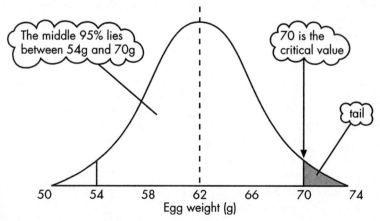

Figure 14.9 Some features of a distribution

TERM	EXPLANATION
Tail	The tail is the small area to the left and to the right of the distribution. The tail can be thought of either as 'the proportion of items falling into this extreme range' or alternatively, 'the probability of a randomly chosen value from the distribution falling into this extreme range'. Both of these definitions basically amount to the same thing. The tail shaded in Figure 14.9 is a measure of the probability that a hen egg, chosen at random, would weigh over 70 g, and this works out at roughly 2½%.
Critical value	The critical value of a distribution of values is a boundary value which marks the divide between the main body of the distribution and the tail. If the value being tested lies outside the critical value (i.e. if it falls into the tail), it is deemed to be 'significantly different' from the mean of the distribution.

Some additional terminology is needed in the discussion of significance testing and the next collection of terms is introduced by means of the legal world of the court-room.

It has already been suggested in this chapter that there are parallels between the ways in which decisions are made in a court of law and in statistics. In this section, the court-room metaphor is extended and developed in order to illustrate some key terms and ideas in statistical decision-making.

Below is a fairly typical summing-up speech which a judge might conceivably make to a jury shortly before they retire to make their decision on the innocence or guilt of the accused. The underlined words parallel certain important statistical ideas which are explained underneath the extract.

> *Members of the jury, you have heard the evidence. I would like to remind you that in our system of justice, an accused person is innocent until proved guilty and so the burden of proof lies with the prosecution, not the defence. I urge you, therefore, to <u>start out with the assumption of innocence</u>. Remember that absolute 100 per cent certainty is never possible to achieve, but you need to <u>be convinced beyond all reasonable doubt</u> of the guilt of the accused before delivering a 'guilty' verdict.*
>
> *You must bear in mind the consequences of a wrong decision. <u>If</u>*

you find the accused 'guilty' when he is actually innocent, then an innocent person is sent to jail. But if you find the accused 'innocent' when he is actually guilty, then a villain goes free. It is the view of this court that better a thousand guilty villains go free than one innocent person is wrongly convicted.

There now follows a 'translation' of these five court-room phrases from 'legal-speak' into 'statistics-speak' in the statistical area that is the subject of this chapter – testing for whether there is a significant difference between two things.

(a) *Starting out with the assumption of innocence:* When testing for a difference in statistics, we start by assuming that there is *no* difference between the values being tested. This is stated as a hypothesis of no difference and is called the null hypothesis (written as H_0). The 'alternative hypothesis' (written as H_1) is the alternative explanation, namely that there *is* a real difference between the values being tested. It is only when the null hypothesis is rejected that the 'alternative hypothesis' can be accepted.

(b) *Being convinced beyond all reasonable doubt:* As the judge remarked, in the real world, 'absolute 100 per cent certainty is never possible to achieve'. In statistics we talk about differences being 'significant to, say, 0.05 (i.e. 5%) or, perhaps, 0.01 (i.e. 1%)'. These figures correspond to the levels of 'reasonable doubt' that the differences might have been just a fluke. The smaller the level of significance, the smaller the 'reasonable doubt' that the differences were due to a fluke result. Thus, a test of significance at the 1% level of significance is a more rigorous test than one at the 5% level of significance.

(c) *Finding the accused 'guilty' when he is actually innocent:* This sort of error is known in statistics as a 'Type 1' error and involves believing that there is a difference between the values when in fact there is no difference. In other words, a Type 1 error is where you accept H_1 when you should have accepted H_0.

(d) *Finding the accused 'innocent' when he is actually guilty:* As you might expect from reading *(c)* above, this is known in statistics as a 'Type 2' error and it occurs where you believe there is no difference between the values when in fact there is a difference – i.e. you accept H_0 when you should have accepted H_1.

These two types of error can be hard to grasp and you may find it helpful to see what they mean in terms of a graph. Figure 14.10 contains two believable boxplots showing the distributions of a large sample of hen eggs and duck eggs, respectively. Suppose, as before, you are given a suspicious egg weighing 72 g and you need to decide whether or not it is a hen egg. The null hypothesis H_0 is that the egg is a hen egg. The alternative hypothesis, H_1, is that it is not a hen egg. You must decide where to draw the critical value. If the suspicious egg falls outside the critical value, then you will reject H_0 and accept H_1. Where you choose to draw the critical value will determine how rigorous the test will be. We will look at two different critical values (marked CV1 and CV2 on the diagram) and how their positioning affects the likelihood of making either a Type 1 or a Type 2 error.

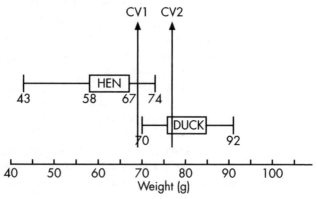

Figure 14.10   Types of error resulting from the positioning of the critical value

First, look at CV1, which is a very low threshold value for a critical value of around 68 g. Clearly there are likely to be quite a large number of genuine hen eggs which are heavier than this value. These eggs will be incorrectly classified as being 'not hen eggs' and therefore the probability of making a 'Type 1' error in this situation is quite high. Next look at

CV2 – a value of around 77 g. This is a much more rigorous threshold value and we are very unlikely to come across any genuine hen eggs that are heavier than this. So, now we are very unlikely to make a 'Type 1' error. However, this opens the door to making the opposing error. There may well be a number of duck eggs (for example, a duck egg weighing, say, 75 g) which we were unable to eliminate from the category 'hen egg'. Now we are making a 'Type 2' error – accepting H_0 when we should have rejected it.

> (e) *Better a thousand guilty villains go free than one inno-cent person is wrongly convicted:* This statement says something about the relative importance of the two types of error described above. If, in a statistical test of significance, you apply a very stringent test (say, applying a low level of significance of 0.001, or 0.1%) this means that you will only accept H_1 if there is an extremely large difference between the values. However, such circumstances make it likely that there may be 'real' differences which are not picked up – in other words, this is a situation where a Type 2 error is likely. However, the other side of the coin is that, under these conditions of a low significance level, you are very unlikely to claim a difference when none exists – i.e. you are unlikely to make a Type 1 error. Correspondingly, in a test of significance with a high level of significance (say 0.1 or 10%), you are likely to make a Type 1 error but unlikely to make a Type 2 error. The message from the judge to statisticians seems to be – 'Set your significance levels low to ensure that you are much more likely to commit a Type 2 error than a Type 1 error. This means that you will miss spotting some significant differences but the differences that you do claim are very likely to be true.'

You may have found some of the terminology and explanations above rather difficult to follow, so don't be afraid to read these five points, (a) to (e) more than once. Table 14.2 summarizes the essential relationship between the two types of error, both in the context of a court-room and in statistical decision-making.

(a) *In the court-room*

		Verdict	
		Not-guilty	*Guilty*
Defendant	*Saint*	A correct decision	Type 1 error
	Villain	Type 2 error	A correct decision

(b) *In statistical decision-making*

		Conclusion	
		No difference (H_0)	*Difference (H_1)*
Reality	*Populations really are the same*	A correct decision	Type 1 error
	Populations really are different	Type 2 error	A correct decision

Table 14.2   Type 1 and Type 2 errors

What is a test of significance?

The terms 'test of significance' and 'hypothesis test' both refer to more or less the same statistical process, of deciding whether or not there is a difference between two values or two sets of results. There are a variety of tests to choose from, the particular choice depending on the nature of the data and the context in which the comparison is being made. However, in general, the procedure for carrying out a test of significance can usually be summarized into four clear stages. These are set out in Figure 14.11 and each stage is also described in turn. In order to give a context for the explanations, we shall again focus on the egg example.

Stage 1   Set up the hypothesis, H_1

The null hypothesis in this example, H_0, would be that 'there is no difference between the weight of the suspicious egg and that of a sample of hen eggs'. So the alternative hypothesis, H_1, is that the suspicious egg is heavier than the sample of hen eggs. Graphically, this might resemble Figure 14.12.

Notice that the wording of H_1 is not that the weight of the suspicious egg is *different* from that of the rest of the eggs, but that its weight is *more*

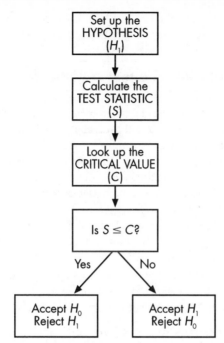

Figure 14.11   Flow chart showing the four main stages of a test of significance

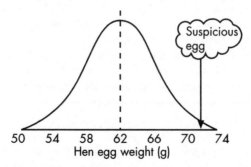

Figure 14.12   Weight of the suspicious egg compared to that of a sample of hen eggs

than these levels. This is an important distinction and it determines whether or not you will need to apply a *one-tailed test* or a *two-tailed test*. This is a one-tailed test because it specifies the direction of the difference (in this case, 'more than'). A test which seeks to establish whether two values are different, but the direction of the difference is not specified, will require a two-tailed test.

Stage 2 Calculate the test statistic, S

The test statistic, S, is some standardized measure of how far the weight of the mystery egg differs from the mean weight of the other eggs against which it is being compared. Which particular form of measure you should choose will depend on certain assumptions about the nature of the data in question. In the final sections, we will look at examples where the data being tested are Normally distributed (giving rise to the 'z test'), and binomially distributed (giving rise to the binomial test). Given that we have been told that the weight of the hen eggs are Normally distributed, we will use a simple test based on our knowledge of the Normal distribution from the previous chapter. We will therefore calculate S as the number of standard deviation units that the weight of the mystery egg is from the mean.

Since the mean is 62 g, an egg of 72 g lies 10 g to the right of the mean. But we know that the standard deviation is 4 g, so the standardized measure $= 10/4$ or 2.5 standard deviations, i.e., $S = 2.5$.

Stage 3 Look up the critical value

A number of key elements of the test must have been decided by this stage. Firstly, you will know what sort of test you are going to use – perhaps a 'z test', a binomial test or a chi-squared test, etc. You will also have decided whether to apply a one-tailed test or a two-tailed test. Finally, you will have selected an appropriate level of significance for the test – perhaps at the 5% or maybe the 1% level. On the basis of all this information, you should now be able to choose the relevant critical value from the appropriate table of critical values. These can be found at the back of most advanced statistics textbooks; and a simplified version of Standardized Normal tables was included in Chapter 13.

Because of the wording of the question – we are testing whether the weight of the suspicious egg is *more than* the mean of the hen eggs, rather than simply *different from* this mean – we are only interested in differences in one direction and will therefore use a one-tailed test. The level of significance is rather arbitrary here, but for the sake of argument, let us choose a 5% level of significance. This means that there will be a 5% tail on the right of the Normal distributional curve. This means a 95% value for $F(z)$. We look up 0.95 backwards in the Standardized

Normal table on page 239 of the previous chapter, which gives a value for z of roughly 1.6. Thus, the critical value, C, is 1.6.

Stage 4 Is $S \leqslant C$?

You must now decide whether or not the test statistic, S, is sufficiently large to warrant believing the alternative hypothesis, H_1. Remember, from Figure 14.10, that we can accept H_1 only if the test statistic, S, lies outside the critical value – i.e. if $S > C$. Otherwise we must accept H_0 and reject H_1.

In this case, the value for S of 2.5 is greater than the critical value of 1.6, so we reject H_0 and accept H_1. In other words, we conclude that the suspicious egg is too heavy to be a hen egg.

Let us now put these ideas into practice with some examples of particular tests of significance. For reasons of space, we will only look, briefly, at two tests, the 'z test' and the binomial test.

The z test

The z test is typically used in a situation where a set of values (a sample) has been taken and we are interested to test whether or not the values are likely to have come from some parent population. Since it would be very time-consuming to test each value in turn against the parent population, the standard procedure for the z test is to find the mean of the values and test whether it is significantly different from the population mean. Here is a typical example of how the test might be applied, based on measures of IQ.

Imagine that a company has marketed an audio-cassette designed to improve people's IQ score. (Perhaps they are required to play the tape every night while they are asleep, or some foolishness!) A sample of four people are tested after using the tape for the prescribed period. Their measured IQ at the end of the course was 100, 112, 92 and 108, respectively, giving a sample mean of 103.

EXERCISE 14.4. Are you convinced?

Are you impressed with these results?

What further information would you like to know in order to make sense of these findings?

Comments follow

There are two important points to investigate here. The first is whether these scores are an improvement at all. It is possible that all four people in the sample started off with IQ scores of over 120 and the subliminal audio-tape actually impaired their brains rather than stimulated them. Secondly, you may not have been impressed with the fact that the sample size was so small. Even if the mean IQ scores were, say, 100 to start with, an improvement of only 3 to the mean is not so impressive. If the sample size had been, say, 2000, then this sort of improvement would have been more noteworthy.

It is this second point that we must now pursue in order to get to the heart of the z test. The key idea to grasp is that, if you were to take a sample from a population and find the sample mean, the larger the sample you take, the closer the sample mean is likely to lie to the population mean. To take an extreme case, if you were to sample the entire population, then the sample mean would exactly equal the population mean with no variation at all. This point has already been demonstrated in Chapter 7 under the heading of sampling variation. However, we shall now be rather more precise about the relationship between the size of a sample and the degree of variation you might expect the sample mean to show.

Let us start by considering some population with a mean denoted by μ (a Greek letter pronounced 'mew') and a standard deviation denoted by σ (pronounced 'sigma').

In general, if you take from this population a series of samples, each of size n, and find the mean of each sample, then the following two interesting results arise.

- ■ The set of sample means, taken together, always form a Normal distribution, whether or not the original population from which they were drawn was Normally distributed.
- ■ The standard deviation of this set of sample means, often called the *standard error* (*SE*) of the means, is equal to

$$\frac{\sigma}{\sqrt{n}}$$

Taken together, these two results, constitute what is known as the 'Central Limit Theorem'. The first of these results is important because it allows us to apply the z test to the mean of any sample, no matter how its underlying population is distributed. Provided we restrict our attention just to sample means, we can correctly use Normal distribution tables in our test of significance.

Notice from the second of these results that the larger the sample size, n, the larger will be $\sqrt{n}$ and therefore the smaller will be the standard error,

$$\frac{\sigma}{\sqrt{n}}$$

With this background, we are now ready to apply the z test to the following two examples.

The z test, Example 1

A sample of 25 people are randomly chosen from a population whose measured IQs are $N(100,225)$.[1] They are each given the same 'brain-enhancing audio-cassette' to use for a prescribed period, after which their IQs are tested. Their mean score after the course is 104. Is there evidence that the course is effective?

Solution

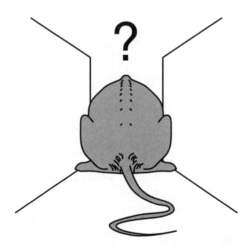

One-tailed test

First we must decide whether to use a one-tailed or a two-tailed test. We will discover this from the way the question is worded. In this case we have been asked to test whether there is 'evidence that the course is

---

[1] As was observed earlier, it was not necessary for the underlying population to have been Normally distributed in order to apply the z test. It just happens that in this case, IQs *are* Normally distributed.

effective'. In other words, we wish to know whether there has been an improvement, not whether there is simply a difference. In cases like this where the direction of the difference is specified, a one-tailed test is used. Next, the level of significance. There is no easy answer to this – the level chosen reflects how rigorous you choose to be. In this case the result is not a life-or-death one, so a 5% level of significance would be appropriate. If the context of the test happened to be something like the health effects of drugs, then a 1% or maybe a 0.1% level of significance might be chosen.

The z test will now be applied using the four stages described earlier.

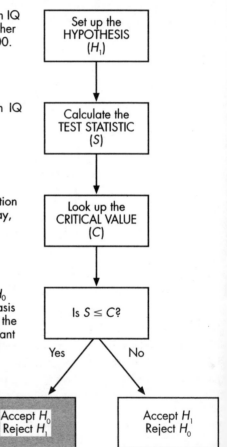

Stage 1: H_1 states that the mean IQ score of 104 is significantly higher than the population mean of 100.

Set up the HYPOTHESIS (H_1)

Stage 2: SD of the population IQ scores is $\sqrt{225} = 15$.

So, $SE = \dfrac{\sigma}{\sqrt{25}} = \dfrac{15}{5} = 3$

$S = \dfrac{104 - 100}{3} = 1\frac{1}{3}$

Calculate the TEST STATISTIC (S)

Stage 3: Using Normal distribution tables, a one-tailed test, and, say, a 5% level of significance, the critical value $C = 1.6$ (approximately)

Look up the CRITICAL VALUE (C)

Stage 4: $S < C$, so we accept H_0 and reject H_1. That is, on the basis of these figures, we accept that the scores do not show any significant improvement over the expected values.

Is $S \leq C$?

Yes   No

Accept H_0
Reject H_1

Accept H_1
Reject H_0

The **z** test, Example 2

Let us now take the same example as before but this time suppose that the results were drawn from a much larger sample.

A sample of 225 people are randomly chosen from a population whose measured IQs are $N(100, 225)$. They are each given the same 'brain-enhancing audio-cassette' to use for a prescribed period, after which their IQs are tested. Their mean score after the course is 104. Is there evidence that the course is effective?

Solution

Stage 1: H_1 states that the mean IQ score of 104 is significantly higher than the population mean of 100.

Stage 2: SD of the population IQ scores is $\sqrt{225} = 15$.

So, $SE = \dfrac{\sigma}{\sqrt{225}} = \dfrac{15}{15} = 1$

$S = \dfrac{104 - 100}{1} = 4$

Stage 3: Using Normal distribution tables, a one-tailed test, and, say, a 5% level of significance, the critical value, $C = 1.6$ (approximately)

Stage 4: $S > C$, so we accept H_1 and reject H_0. That is, on the basis of these figures, we accept that the scores do seem to show significant improvement over the expected values.

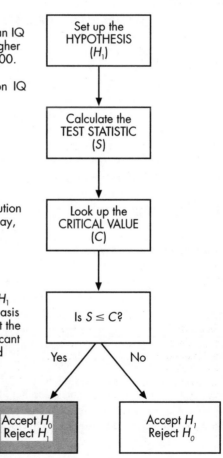

Set up the HYPOTHESIS (H_1)

Calculate the TEST STATISTIC (S)

Look up the CRITICAL VALUE (C)

Is $S \leq C$?

Yes   No

Accept H_0
Reject H_1

Accept H_1
Reject H_0

Now here a few examples to try for yourself.

EXERCISE 14.5. Applying the z test

a) *Assume that hen eggs are distributed with a mean of 62 g and a standard deviation of 4 g. A random sample of hens are given a specially prepared feed and 4 of their eggs are subsequently weighed, giving a mean of 66 g. Is there evidence that the feed is effective?*

b) *A machine weighs bags of crisps with a mean weight of 34 g and a standard deviation of 4 g. The quality control checker is concerned that the machine may be faulty. A sample of 25 bags are weighed and they show a mean weight of 33 g. Is there evidence that the machine is faulty?*

c) *A large number of leaves of species A were measured and their lengths were found to have a mean of 64 mm and a standard deviation of 8 mm. A particular bush is discovered with leaves that are visually similar to species A. A sample of 100 leaves from the bush were collected and they were shown to have a mean length of 61 mm. Do you think the bush is species A?*

Comments on page 281

Before leaving the z test, it should be stressed that we have only looked at a narrow set of questions here. In each example you have been told the mean and standard deviation of the underlying population and asked to test whether or not a sample mean is likely to have come from that population. There are a variety of other situations where a similar test would be useful, for example where data have been collected from two samples and you wish to know whether they are likely to have come from the same population. There is not space here to pursue such questions, but you should be aware that the range of tests similar to the one described above is extensive. The purpose of this chapter is simply to give a flavour of one or two simple tests.

The binomial test

The binomial model was introduced in the last chapter as an important probability model. There we described a binomial probability distribution as resulting from a series of events, each of which would yield only one of two possible outcomes. These outcomes we called 'success' and 'failure'. In this section we will look at just two examples where a binomial test would be appropriate.

The binomial test, Example 1

A coin is tossed six times and results in a total of five 'heads' and one 'tail'. Is this sufficient to suggest that the coin is biased towards heads?

Solution

In order to make a start on this question, we need to examine the underlying binomial model. It can be represented by the binomial distribution with the letter 'p' representing the probability of tossing 'heads' at each toss, 'q' representing the probability of tossing 'tails' at each toss, and n corresponding to the number of tosses (in this case, $n = 6$).

The full binomial expansion is as follows:

$$(p + q)^6 = p^6 + 6p^5q + 15p^4q^2 + 20p^3q^3 + 15p^2q^4 + 6pq^5 + q^6$$

NUMBER OF HEADS	TERM	VALUE WHERE $p = q = \frac{1}{2}$	DECIMAL VALUE
6	p^6	$(\frac{1}{2})^6$	0.016
5	$6p^5q$	$6 \times (\frac{1}{2})^6$	0.094
4	$15p^4q^2$	$15 \times (\frac{1}{2})^6$	0.234
3	$20p^3q^3$	$20 \times (\frac{1}{2})^6$	0.313
2	$15p^2q^4$	$15 \times (\frac{1}{2})^6$	0.234
1	$6pq^5$	$6 \times (\frac{1}{2})^6$	0.094
0	q^6	$(\frac{1}{2})^6$	0.016
			TOTAL 1.001[1]

Table 14.3   Listing the separate probabilities of each term in the expansion of $(p + q)^6$

[1]Note that these probabilities should add to 1. The slight error here is due to rounding.

From this table we can deduce just how likely it would be to produce various outcomes from chance alone by tossing a fair coin. For example:

- The probability of tossing six 'heads' = 0.016
- The probability of tossing at least
  five 'heads' = 0.016 + 0.094
  = 0.110

and so on.

Stage 1: H_0 states that the coin is fair. H_1 states that the probability of tossing 'heads', p, is significantly larger than 0.5.

Stage 2: The test statistic, S, is simply the number of 'heads', which in this case is 5. So $S = 5$.

Stage 3: Using a one-tailed test, and, say, a 5% level of significance, the critical value, C is found by identifying the number of 'heads' required to produce a tail that is *just more than 5%*. As has already been calculated, '5 or more heads' produces a tail of 0.11, which is just too large, so $C = 5$.

Stage 4: $S = C$, so we accept H_0 and reject H_1. That is, on the basis of these figures, we do not accept that the coin is biased towards 'heads'.

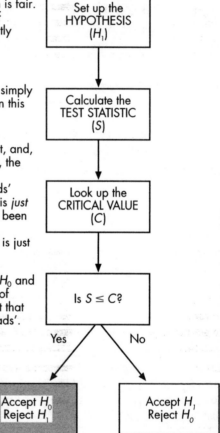

We also need to decide whether to use a one-tailed or a two-tailed test.

Because the question is worded in such a way as to specify a particular direction of bias (in this case, 'towards heads') we will adopt a one-tailed test. If it had simply asked whether there was evidence of bias, then a two-tailed test would have been used.

Having completed the preliminaries, we can now begin the formal test. As with the z test, we shall use the familiar four-stage significance testing model.

The binomial test, Example 2

Educational psychologists are interested in how and under what circumstances learning takes place. One classic experiment carried out by Clark Hull was designed to investigate whether or not a rat was capable of putting together two separate pieces of information in a logical way. Hull set out a maze as shown in Figure 14.13.

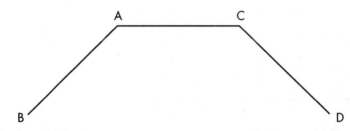

Figure 14.13

A rat learned to run from A to B to get a small reward, and from A to C to get an equally small reward. Finally it learned to run from C to D to get a much bigger reward. Next, the rat is placed at position A. If it has learned to put together the previous learning logically, it should choose to run to C rather than B, as this will ensure that it will be able to go on to get the large reward going from C to D. If no learning has taken place, then there is a 50:50 chance of it taking either route AB or AC.

Let us suppose that ten rats were trained in the manner described above and each one was placed, separately and in turn, at point A. In the event, seven rats ran to C and only three ran to B. Is there evidence to suggest that learning has occurred?

Solution

The probability model against which these results are to be compared is the binomial distribution. Let 'p' correspond to the probability that a rat takes the 'intelligent' route along AC and 'q' be the probability that a rat takes the 'unintelligent' route along AB. Let the number of trials, $n = 10$, and $p = q = 0.5$. In other words, we want to test:

H_0: $p = 0.5$ i.e. no learning has taken place

H_1: $p > 0.5$ i.e. learning has taken place.

Since we are testing whether the rats are performing *better* than expected from chance alone (i.e. not simply *different* from chance alone) then a one-tailed test must be used.

We start by looking at the tail of the binomial distribution. For example, the probability of scoring 8 or more successes from this distribution is

$P(X = 8) + P(X = 9) + P(X = 10)$.

The probability of scoring 9 or more successes from this distribution is

$P(X = 9) + P(X = 10)$.

The calculation of these probabilities is found from evaluating the last three (or two) terms from the binomial expansion of $(q + p)^{10}$. The full binomial expansion of p^{10} is as follows.

$(p + q)^{10} = p^{10} + 10p^9q + 45p^8q^2 + 120p^7q^3 + 210p^6q^4 + 252p^5q^5 + 210p^4q^6 + 120p^3q^7 + 45p^2q^8 + 10pq^9 + q^{10}$

This gives:

$P(X = 8) = 45(\tfrac{1}{2})^{10} = 0.044$

$P(X = 9) = 10(\tfrac{1}{2})^{10} = 0.010$

$P(X = 10) = (\tfrac{1}{2})^{10} = 0.001$

Taking these together, we get the areas of the tails for $X = 10$, $X \geqslant 9$ and $X \geqslant 8$, as follows.

$P(X = 10) = 0.001$

$P(X \geqslant 9) = 0.010 + 0.001 = 0.011$

$P(X \geqslant 8) = 0.010 + 0.001 + 0.044 = 0.055$

We can now apply our four-stage model and answer the original question.

Stage 1: H_1 states that the probability, p, of a rat choosing route AC is significantly larger than 0.5.

Stage 2: The test statistic, S, is simply the number of 'successes', which in this case is 7. So $S = 7$.

Stage 3: Using a one-tailed test, and, say, a 5% level of significance, the critical value, C, is found by identifying the number of successes required to produce a tail that is *just more than 5%*. As has already been calculated '9 or more successes' gives a tail of 0.016, while '8 or more successes' produces a tail of 0.055, which is just larger than 5%. So $C = 8$.

Stage 4: $S < C$, so we accept H_0 and reject H_1. That is, on the basis of these figures, we cannot claim that the rats have shown learning of the type described.

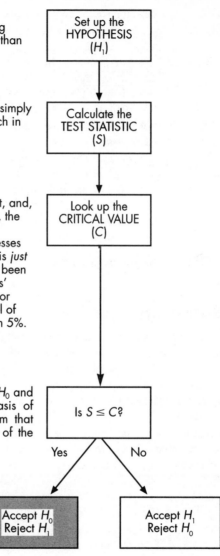

You should now try some binomial tests for yourself, so have a go at Exercise 14.6.

EXERCISE 14.6. Applying the binomial test

a) *The rat test that was described above was repeated, and this time 9 out of the 10 rats chose to travel along the 'intelligent' route, AC. Does this provide evidence that the rats have shown learning of the type described?*

b) *In a multiple choice examination, students must choose from a choice of four possible answers on each question. Out of six questions, a student got five correct. The null hypothesis is that the student is guessing. Test the null hypothesis. (Hint: if there is a one in four chance of guessing correctly, then set p = $\frac{1}{4}$)*

Comments on page 282

Summary

This chapter has provided a brief whistle-stop tour of a major area of statistics called tests of significance, otherwise known as hypothesis testing. The early sections were an attempt to lay down some of the principles and terminology of this formal approach to deciding on differences. A key concept here is that, due to natural variation, you would expect differences to occur by chance. The essential principle underlying significance testing is that the difference being investigated must be *sufficiently large* to warrant reaching the conclusion that there is a 'real' difference, and unlikely to be one due to chance alone. Two important types of error were explained – known as Type 1 and Type 2 error. A Type 1 error occurs where a difference is claimed when none actually exists. A Type 2 error is the converse of this – i.e. it occurs when there really is a difference between the things being tested but the test fails to pick this up.

The chapter ended with a brief look at two common tests, the z test, and the binomial test.

Comments on exercises

Exercise 14.5

(a) *Stage 1:* H_1 states that the mean egg weight of 66 g is significantly heavier than the population mean of 62 g.

Stage 2: SD of the population weight = 4g.

So, $SE = \dfrac{\sigma}{\sqrt{4}} = \dfrac{4}{2} = 2$

$S = \dfrac{66 - 62}{2} = 2$

Stage 3: Using Normal distribution tables, a one-tailed test (so chosen because the question asks whether there is an improvement, not merely a difference), and, say, a 5% level of significance, the critical value, $C = 1.6$ (approximately)

Stage 4: $S > C$, so we accept H_1 and reject H_0. Thus, on the basis of these figures, we accept that the egg weights do show a significant improvement over the expected values, suggesting that the feed is effective.

(b) *Stage 1:* H_1 states that the mean weight of 33 g is significantly different from the population mean of 34 g.

Stage 2: SD of the population weight = 4 g.

So, $SE = \dfrac{\sigma}{\sqrt{25}} = \dfrac{4}{5} = 0.8$

$S = \dfrac{33 - 34}{0.8} = -1.25$

(Note that the minus sign can be ignored here. We are interested only in the absolute value of S. So $S = 1.25$.)

Stage 3: Using Normal distribution tables, a two-tailed test (so chosen because the question asks whether there is any difference, not specifically either an increase or a decrease), and, say, a 5% level of significance, the critical value, $C = 2.0$ (approximately).

Stage 4: $S < C$, so we accept H_0 and reject H_1. That is, on the basis of these figures, we accept that the sample of bags are not significantly different in weight from the expected values, suggesting that the machine is not faulty.

(c) *Stage 1:* H_1 states that the mean length of 61 mm is significantly different from the population mean of 64 mm.

Stage 2: SD of the population length = 8 mm.

So, $SE = \dfrac{\sigma}{\sqrt{100}} = \dfrac{8}{10} = 0.8$

$S = \dfrac{61 - 64}{0.8} = -3.75.$

Again we can ignore the minus sign as we are only interested in the absolute value of S. So, $S = 3.75$.

Stage 3: Using Normal distribution tables, a two-tailed test (so chosen because the question asks whether there is a difference, not specifically either an increase or a decrease), and, say, a 5% level of significance, the critical value, $C = 2.0$ (approximately).

Stage 4: $S > C$, so we accept H_1 and reject H_0. Thus, on the basis of these figures, we accept that the sample of leaves from the mystery bush are significantly different in length from the expected values and are therefore unlikely to be species A.

Exercise 14.6 Applying the binomial test

(a) *Stage 1:* H_0 states that a rat is equally likely to take either route. H_1 states that the probability, p, of a rat choosing route AC is significantly larger than 0.5.

Stage 2: The test statistic, S, is the number of 'successes', which in this case is 9. So $S = 9$.

Stage 3: Using a one-tailed test, and, say, a 5% level of significance, the critical value, C, is found by identifying the number of 'successes' required to produce a tail that is just more than 5%. As before, $C = 8$.

Stage 4: $S > C$, so we accept H_1 and reject H_0. Therefore, on the basis of these figures, we can claim that the rats have shown learning of the type described.

(b) *Stage 1:* H_1 states that the probability, p, of choosing the correct answer is significantly larger than the probability of getting it right by guess-work alone, which is 0.25.

Stage 2: The test statistic, S is the number of 'successes', which in this case is 5. So $S = 5$.

Stage 3: Using a one-tailed test, and, say, a 5% level of significance, the critical value, C, can be calculated from Table 14.4.

NUMBER CORRECT	TERM	VALUE WHERE $p = 0.25$, $q = 0.75$	DECIMAL VALUE (to 3 d.p.)
6	p^6	$(\frac{1}{4})^6$	0.000
5	$6p^5q$	$6 \times (\frac{1}{4})^5(\frac{3}{4})$	0.001
4	$15p^4q^2$	$15 \times (\frac{1}{4})^4(\frac{3}{4})^2$	0.033
3	$20p^3q^3$	$20 \times (\frac{1}{4})^3(\frac{3}{4})^3$	0.132
2	$15p^2q^4$	$15 \times (\frac{1}{4})^2(\frac{3}{4})^4$	0.300
1	$6pq^5$	$6 \times (\frac{1}{4})(\frac{3}{4})^5$	0.356
0	q^6	$(\frac{3}{4})^6$	0.178
			TOTAL 1.000

Table 14.4

From the table we get

$P(X = 6) = 0.000$

$P(X \geqslant 5) = 0.000 + 0.001 = 0.001$

$P(X \geqslant 4) = 0.000 + 0.001 + 0.033 = 0.034$

$P(X \geqslant 3) = 0.000 + 0.001 + 0.033 + 0.132 = 0.166$

The critical value, C, is found by identifying the number of 'successes' required to produce a tail that is just more than 5%. In this case, $P(X \geq 4)$ is just too small, so we choose $P(X \geq 3)$. Thus, $C = 3$.

Stage 4: $S > C$, so we accept H_1 and reject H_0. That is, on the basis of these figures, we can claim that the student is unlikely to have guessed the answers.

FURTHER READING

Erickson, BH and Nosanchuk, TA, (1988) *Understanding Data*, Open University Press

Francis, A (1988) *Advanced Level Statistics*, Stanley Thornes (Publishers) Ltd.

Graham, A (1991) *Investigating Statistics, a Beginner's Guide*, Hodder and Stoughton

Graham, A (1995) *Teach Yourself Basic Maths*, Hodder and Stoughton

Huff, D (1973) *How to Lie with Statistics*, Penguin

Huff, D (1978) *How to Take a Chance*, Pelican

Marsh, C (1988) *Exploring Data*, Polity Press

Moore, DS and McCabe, GP (1989) *Introduction to the Practice of Statistics*, WH Freeman and Company

Moroney, MJ (1971) *Facts from Figures*, Pelican

Paulos, JA (1988) *Innumeracy*, vi

Reichmann, WJ (1978) *Use and Abuse of Statistics*, Pelican

Useful sources of data

Central Statistical Office (1999*) *Annual Abstract of Statistics*, HMSO

Central Statistical Office (1999*) *Regional Trends*, HMSO

Central Statistical Office (1999*) *Social Trends*, HMSO

Department of Employment (1999*) *Family Expenditure Survey*, HMSO

Department of Employment (1999*) *New Earnings Review*, HMSO

Equal Opportunities Commission (1999*) *Women and Men in Britain: a Research Profile*, HMSO

Jowell, R, Witherspoon, S and Brooks, L (1999*) *British Social Attitudes*, Gower Office of Population Censuses and Surveys, *General Household Survey*, (1999*) HMSO

* These publications are produced annually or on a regular basis. Look out for the most recent edition.

APPENDIX

Choosing the right statistical technique

The flow chart below summarizes the main stages in statistical work.
Relevant chapter numbers are given in brackets.

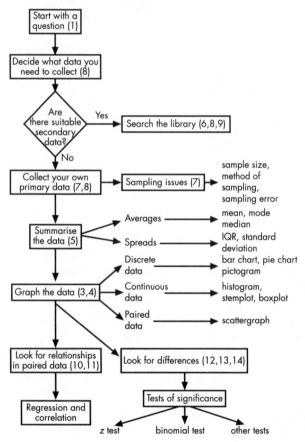

INDEX